The Dogs of
Windcutter
Down

Also by David Kennard

A Shepherd's Watch
The Year of the Working Sheepdog (DVD)

David Kennard

The Dogs of Windcutter Down

One Shepherd's Struggle for Survival

headline

First published in 2005
by HEADLINE BOOK PUBLISHING

1

Cataloguing in Publication Data is available from the British Library

0 7553 1256 2

Typeset in Scala by Avon DataSet Ltd,
Bidford-on-Avon, Warwickshire

Designed by Janette Revill

Printed and bound in Great Britain by
Mackays of Chatham plc, Chatham, Kent

Headline's policy is to use papers that are natural, renewable and recyclable
products and made from wood grown in sustainable forests. The logging
and manufacturing processes are expected to conform to the environmental
regulations of the country of origin.

HEADLINE BOOK PUBLISHING
A division of Hodder Headline
338 Euston Road
London NW1 3BH

www.headline.co.uk
www.hodderheadline.com

For the children who brighten Borough Farm:
Kate, Jemima, Callum, Ros, Frances, Anna,
Leanne and Frances L.
But most of all Clare, Laura and Nick.

Contents

Contents

CHAPTER ONE

It Never Rains

A wicked blast of Atlantic wind wrenched the gate from my
hand and sent it crashing shut. The horizontal rain was
relentless, stinging my face and forcing my eyes to narrow
slits. Violent storms had been blowing in for over a week now,
and the whole North Devon landscape looked battered and
sodden. A line of ash trees that ran along the stone wall had
been bent almost double. Small branches were strewn far and
wide, some half immersed in what days ago had been puddles
but had now grown into small lakes.

The front wheel of the Land Rover stood in six inches of

water inside the gate near the roadside hedge. Not a day for driving further into the field, I decided. The last thing I wanted to do was get stuck in this weather.

The reason for my visit to this corner of Town Farm, the land I rented in the coastal village of Mortehoe, lay a hundred yards away. A small group of lambs stood there, soaked and miserable, their backs to the wind, sheltering in the lee of a stone wall and some scattered clumps of gorse. The storms had begun to leave their mark on this flock of two hundred and fifty lambs. In the past four days I'd found three dead. Two more were so sick I'd had to take them on the short drive inland to the sheds at home on Borough Farm where they were undergoing treatment. I was almost certain of the cause: pasteurella, the biggest single killer of sheep in the country. As I made my way round to the back of the Land Rover, I was quietly dreading discovering more victims during the latest, daily check of the flock.

No sooner had I hauled open the door of the Land Rover than two of my sheepdogs had leapt out. Gail and Swift were eager for an outing despite the foulness of the morning. The third occupant of the rear compartment wasn't quite so keen, however. Greg, my oldest dog, stayed put, peering out at the grey, rain-lashed world with little enthusiasm. I knew that look. He would have joined the rest of us if I'd asked, but he was just as happy to stay put.

'Good decision, Greg,' I muttered to myself, as the wind gusted once more, ripping the door from my hand and hammering it shut. I fastened the waterproof coat tight around my neck, and, with dogs at heel, set off towards the scattered flock.

Pasteurella is one of those bugs that is always present but a sheep can normally fend off. It only causes a real problem when the sheep's immunity is low. There are all sorts of factors that can reduce a sheep's ability to fight diseases; living in the face of a seemingly endless storm was one of them. At more than seven months of age, each of these lambs now possessed a thick coat easily capable of repelling the wind and rain, but coping with the constant bombardment from the weather was a relentless physical challenge and it had left them vulnerable and stressed.

The idea that sheep suffer stress may be difficult for many people to accept – after all, they are, for the most part, not the most intelligent and sensitive of creatures – but it is a very real problem, in its way every bit as harmful as a physical ailment.

Spotting the early signs of impending illness wasn't easy either. Generally a shepherd can identify a sickly sheep by its behaviour. It might hang away from the rest of the flock or refuse to be moved when being gathered, for instance. In these conditions, however, it was hard to tell anything. Today no one sheep was distinguishable from the rest. They all stood sheltering from the buffeting winds, sullen and drained of all enthusiasm for life. If I was going to have any chance of detecting any new cases of disease, then I must persuade them to move a little.

Gail, now three years old, stood at my side, her eyes nearly shut facing the stinging rain, her coat parted to the skin on the windward side. At the sound of a quiet 'Away, Gail' from me, she moved purposefully off into the wind, and began turning small groups of lambs into the middle of the field.

Swift, aged six, is Gail's mother. While her daughter worked one side of the field, she stood patiently with me, her gaze fixed on the other side where she was anticipating I would soon send her. As ever, she was focused completely on work. It was as though she hadn't noticed the weather. With a 'Come bye, Swift' I sent her off and she was soon working in tandem with Gail. A reluctant straggle of lambs began grouping together. I started walking towards them, not wanting them to be driven any further than necessary.

A good stockman, it's said, will look for trouble, but hope to see nothing. In the event, nothing jumped out at me as being unusual. Part of me felt relieved, yet another part of me was counselling caution. I was just about to walk around the rest of the field, when I caught sight of two crows sweeping into the air from a large gorse clump. My heart sank as I changed direction and headed towards it. There could be only one reason for the crows' interest.

In the sheep-worn hollow of the centre of the bush, lay a new victim. It had a bloated belly and there was blood-stained froth bubbling from its nose, strong indications of pasteurella. Inwardly I'd been hoping the problem would pass over, but this now snuffed out any flickering sense of optimism. Wearily, I dragged the carcass back and loaded it onto the trailer. Swift and Gail had rejoined me and they clambered into the back of the Land Rover to be reunited with Greg, grateful, no doubt, to find the shelter of the cab once more. As I climbed back in and shut the door, beads of rain trickled down my neck. I unbuttoned my coat, and wiped my face dry on a dirty, damp jumper that had lain on the floor of the truck for weeks.

'It never rains, but it pours,' I mumbled to myself, as the wipers flicked back and forth at double speed, all the way back to the farm.

In the spring of this year, foot and mouth disease had struck mainland UK for the first time in over thirty years. The outbreak's impact was immediate and almost overwhelming. For my wife Debbie and our children Clare, Laura and Nick, the ever-present threat of the disease reaching Borough Farm became the complete be-all-and-end-all of our lives. For six months from the middle of February, our days became a seemingly endless cycle of spraying cars and washing wellies in disinfectant.

The daily checks on the local radio news and the Ministry website became a ritual filled with dread. The loss of the stock of many friends and acquaintances was reported with sickening regularity. Worse still, through much of March and April the direction of the spread seemed to indicate that it was 'when' not 'if' the invisible curse would arrive at Borough Farm.

Yet we were spared, and for that I believe we had to thank the farmers only a few miles away who were not so lucky. It must be one of the most difficult calls to make, to phone the Ministry to tell them that you believe your stock has become infected with foot and mouth. It effectively signs their death warrant. One friend made that call and found himself unable to speak. After he eventually hung up, the number was traced and his stock destroyed.

The images of burning pyres of livestock that filled the evening news bulletins showed all too graphically the devastating impact on the countryside's animal population, but the human cost of the disease was invisible, and incalculable. Although our flock had survived, even now in November, eight months on from the initial outbreak, the effects were still reverberating.

In the spring, five days after the disease had been declared, all livestock movements had been banned and abattoirs closed down, leaving stock effectively unsaleable. I was on the verge of lambing, and thus the arrival of a new stock of a thousand or so lambs, but, more frustratingly, three hundred of my previous year's lambs were still on the farm, where they were being plied with expensive feed. They should have been sold over those next few weeks, but by the time the restrictions were slackened, several weeks later, so that movements to certain slaughterhouses were permitted under licence, there was a surfeit, so we faced a buyers' market – and with it a hopelessly poor price.

Because they'd remained on the farm past the point where they'd cut their adult teeth, the price of these older lambs was going to be drastically reduced anyway. Even in normal trading conditions, this is the point at which the processing industry deems they halve in value, but, with my regular outlet still closed, I didn't even have a buyer with whom I could negotiate something approaching a fair price, and I'd been forced to trawl around the few businesses that might have been willing to risk buying stock from an area that had been so close to an outbreak. I'll never forget the phone conversation I had with one buyer.

'There's a bloke on the phone with a hundred lambs to sell . . . How much shall we offer him?' I heard one lady call to her colleague, not even bothering to cover the receiver with her hand.

The price I eventually got barely covered the cost of feeding them over the previous prolonged winter months.

Now, in November, with six hundred of this year's lambs still to sell and prices still flat, the farm was once again facing a significant loss.

When Debbie and I had moved to Borough Farm, nine years earlier, we'd done so with our eyes wide open. We'd accepted that the financial returns from a few hundred sheep were never likely to be that great, but we felt that the job and the way of life it gave us and the children would always be more than ample compensation. And so it had proved – until now.

For the first time, the balance had tipped in the other direction. As the winter approached, the picture was bleak: the farm's finances were in free-fall and for the life of me I couldn't see the turning point. We couldn't carry on like this much longer. Which was why, after all we'd gone through this year, it seemed such a cruel twist of fate, to be facing an outbreak of pasteurella.

———

The Land Rover bumped down the lane into Borough Farm. Debbie wandered across the yard, clad in a waterproof coat, her dark hair catching a sudden gust, as she carried a bucket towards the feed-shed.

'Well?' she said, scraping wayward strands of hair from her face and peering into the trailer as I climbed out.

She soon spotted my rather grisly cargo. 'At least it's only one today,' she said, trying to put a brave face on things.

She knew it wasn't going to lighten my mood.

On the way home I'd been to see the vet, who'd wasted little time in confirming my fears.

'See the colour of the lung,' he'd said, as he scooped the salmon pink organ from the carcass he'd just opened up. 'That area of dark red tissue can't absorb any oxygen. It's certainly pneumonia, and ten to one it's brought on by pasteurella.'

I explained this to Debbie in a dour monotone.

'What does he reckon you should do?' she said, trying to steer the conversation into positive territory.

If only there had been a straightforward answer. The lambs had all received one vaccination against the pasteurella bug. The problem was they wouldn't be fully protected until they received a second dose in two weeks' time.

'Moving them to a more sheltered field would help,' I said,

'but that's a risk as well because the stress of the move might finish a few more off. The best thing is just to hope that the weather improves a little, and give them the second jab a week early.'

To be honest, neither of these options seemed satisfactory, but the only other course of action was to treat every lamb with a dose of penicillin, something that went against all my instincts. Avoiding the introduction of antibiotics into the food chain is one of the fundamentals of good husbandry.

'Perhaps we'll leave it one more day,' I said. 'If there are any more problems tomorrow I'll have to go through them with penicillin.'

Debbie could see that there was little chance of lightening my mood.

'Come on, it's lunchtime; if you come in now we can catch the weather forecast.'

A flurry of leaves rustled across the yard as we walked towards the house, blown from the recently planted saplings on the bank below. Greg and Swift ran along ahead and stood expectantly at the door.

Ordinarily I keep all five dogs kennelled outside, but in recent months Greg had begun to spend more and more time in the house. I didn't have a problem with this; to me it was a right he had more than earned during his years of sterling service. The children loved having him indoors, and also had a soft spot for Swift, the other senior citizen of the kennel. For obvious reasons, however, Debbie wasn't always so keen to have the dogs inside with us.

'They're all wet,' she protested today, clearly worried at the prospect of wet, muddy tails being wagged over her walls and sofa, but she soon backtracked. 'All right, but if they come in they'll have to stay in the dungeon,' she relented, referring to the small boot room at the lower level of the house.

'Stop there, Greg, Swift,' I said in a sharp voice, knowing full well that they would soon skulk up the stairs to join us.

The radio didn't provide much cause for optimism. The deep low-pressure area that had settled to the north-west showed little sign of moving away. Gales would continue for at least the next day, and tomorrow it might be cold enough for sleet. 'Great,' I muttered, turning to the sandwich that had appeared on a plate in front of me.

A moment later Swift sneaked into the kitchen without Debbie spotting her, and quietly took up residence by the Aga. Stretching out and shutting her eyes, she looked a picture of contentment.

I was just about to head back out, when the phone rang, startling Swift from her semi-slumber.

'Mr Kennard?'

My mood wasn't exactly lifted by the voice on the other end of the line. It was a Ministry vet, following up an application that I had made to send some lambs away to 'keep' on the richer pastures of a nearby dairy farm for the winter. This had been the way of things for eight months now: any movement of livestock, no matter how small and seemingly unimportant, had to be cleared by the Ministry who issued an official form that who sent to the Trading Standards Office. It wasn't just a matter of them giving a quick OK, either.

'We need to do some blood-testing on your sheep before we can approve your application to make a movement,' the vet explained.

'How many do you need to test?' I asked with some trepidation.

'About 80 per cent of the flock. How many would that be?'

'Around eight hundred, but they're spread over seven flocks, in four places,' I explained, half hoping that she might decide that they didn't all need to be tested. Gathering such a large number of scattered sheep in so many locations was going to be a long job.

'That's fine. We'll come tomorrow morning. Would nine o'clock be all right?'

———

The visitors that gathered in the field looked better suited for a day trip to the moon. Each member of the Ministry team was clad from head to toe in a white space suit.

The weatherman hadn't been wrong. The relentless north-east gale brought with it sheets of face-numbing sleet, and as the inspectors stood huddled around a couple of hundred sheep, gathered in a hastily constructed pen, their hoods flapped in the ferocious wind. I could almost hear teeth chattering. There wasn't a smile to be seen.

I wasn't exactly cock-a-hoop to be going through this rigmarole either. Long before foot and mouth I'd grown used to having visits by inspectors from assorted government agencies and environmental associations, but a part of me always resented having my farm taken over by the bureaucracy that these inspectors represented. Part of the attraction of being a farmer, for me, at least, has always been the freedom it provides. To have officials dictating what I could and couldn't do with my stock went against the grain.

Yet, in this case, I was all too aware of the bigger picture. There were more important issues at stake than just my farm. The team and their leader, a vet called Julia, had my sympathies.

First of all they were only doing their job. These blood tests allowed the Ministry to make sure foot and mouth hadn't spread unnoticed beyond the known areas of infection. With the outbreak under control, the last thing anyone needed now was a farmer unwittingly making a movement of livestock that were carrying the disease.

Also, it took a certain level of hardiness to do this job. If the last nine months had been difficult for farmers, it had, if anything, been even worse for the vets at the front line. Day in, day out they had been forced to make monumental and heartbreaking decisions, often ordering the destruction of hundreds, sometimes thousands, of perfectly fit animals in the attempt to form a 'firebreak'.

For many vets it had been too much. In the months following the outbreak, many had left the profession altogether; many more had opted for the easier life of a warm surgery filled with a stream of small, household pets.

Typically today's four-man testing team included a couple of agricultural lab technicians whose jobs had been put on hold because of the epidemic, and so they simply needed a new way of earning a living. Like everyone else they longed for the day when the outbreak was consigned to history and they could return to the life they'd had before its arrival. Sadly that was still some way off.

At least the blood-testing procedure was speedy. I drove a steady stream of sheep into a small pen, and one member of the team held each animal while a second inserted the needle into a vein, filling a test-tube in a matter of seconds. Crate after crate of tubes were carefully labelled, and stored away, before the last of the sheep were let from the pen.

'We're going to have to get a move on if we're going to get all these flocks tested before dark,' Julia said, glancing at her watch, then squinting towards the heavy grey sky.

To avoid the lengthy process of disinfecting the Ministry vehicles between each field, all five of us plus three wet dogs clambered into my Land Rover, waterproofs dripping and clothes damp to the skin. The windows steamed up with thick condensation as we headed for another flock, and another Siberian blast.

Despite the conditions and the monotony of her task, Julia was much more talkative than the previous inspectors I'd encountered. She'd been in the thick of things during the spring, and knew the story of the area's encounter with foot and mouth more intimately than me.

'It didn't look as if the disease could be stopped in March and April,' she said, shaking her head slowly, as we passed fields dotted with livestock. 'At one point we thought the outbreak would wipe out most of the farms in North Devon.'

As we passed over the brow of a hill, a rare shaft of sunlight lit up the landscape and lifted the mood inside the Land Rover too.

'Still, hopefully it's coming to an end now,' she said.

'Do you think so?' I said, seizing on her optimism.

'Yes, there's only been one positive test in Devon over the past three months. Long distance movements might be different, but I reckon the rules on routine blood-testing for small movements like yours will be lifted soon.'

By the time the last flock was finished, the light was fading. As we dealt with the final pieces of paperwork, Julia shook my hand and wished me well.

'We'll get back to you with the results as soon as we can – within two weeks, with a bit of luck. Then I hope we won't have to see you again in these circumstances, Mr Kennard,' she said, climbing back into the car in the yard.

This had been the third time that I had undergone the blood-testing ritual. The prospect of it being the last was very, very appealing.

———

While two-thirds of my flock remain on Borough Farm, the other third resides on the two large swathes of pasture I rent in and around Mortehoe: Town Farm, near the village, and Morte Point, a spectacular rocky outcrop that stretches a mile into the sea, immediately to the west. As the flocks at Mortehoe had been kept completely separate from those at Borough Farm, they had been excused the day's testing. They'd remained on my mind nevertheless. I still had no idea as to whether the pasteurella outbreak would spread further.

With the Ministry team gone, I had a few minutes to check them. The familiar feeling of dread crept over me as the fields loomed into view. At least the rain had relented, and the wind seemed to be dropping off. Now a damp penetrating mist was being blown in from the sea, threatening to rob what was left of the daylight.

As I went through the usual routine the lambs came towards me with rather more enthusiasm than they had done in the past days. Perhaps they'd sensed that the worst of the weather had passed. Checking the gathered sheep, I could detect no obvious problems. A good sign, I thought. Tentatively I toured the rest of the field, peering into the bushes in the half dark. I couldn't be certain, but there didn't seem to be any more fatalities. I heaved a quiet sigh of relief.

Back at the farm with the light now gone, I checked on the two lambs that were undergoing treatment in the sheep shed. One stood in the corner of the pen, tentatively nibbling on some sweet-smelling hay, a marked improvement on the evening before. It looked as if it had turned the corner.

The other patient's progress was much less encouraging. The lamb stood with his head hung low and ears down. He was drawing heavily on each gurgling breath and a thick drool ran from his mouth. I'd seen enough sick sheep over nearly twenty years of shepherding to know that his suffering was in vain. He wasn't going to recover.

However much I dislike having to destroy livestock, there are times when it is the only option. I left the shed, only to return, shotgun in hand. A moment later the pasteurella had claimed another victim.

I walked slowly back to the house. Even the dogs recoiled from me. Collies seem to fear the gun instinctively and all five dogs disappeared back into their kennels. I reassured them and Greg came out to give me a brief wag of his tail, but he soon retreated again to join the others. Each of them looked tired and bedraggled, drained by the days of miserable weather. Right now I knew precisely how they felt.

CHAPTER TWO
Unwelcome Visitors

As the rainstorms of November passed, the northerly winds cut up the bare lifeless Borough Valley. The trees had been stripped clean, their silhouettes now stark and skeletal against the slate-like sky. Only the rooks passing overhead hinted at the myriad varieties of wildlife surviving the hardships of winter in the thick woodland below.

At least the rooks could look further afield for their sustenance. Flying high, buffeted by the relentless gusts, they called to each other as they headed off towards some distant corn field, where, doubtless, they were about to make

THE DOGS OF WINDCUTTER DOWN

themselves unpopular by pulling up the newly germinated corn, in search of the grubs living on the roots.

I was standing on an outcrop, watching the rooks wheeling away into the distance when a sudden nudge in the back left me momentarily off balance. I spun round to see the familiar face of Ernie, the youngest of my dogs.

'Ernie, off!' I scolded him, half laughing.

Now nearly three, Ernie was something of an enigma. On the one hand, he was far and away the most difficult of my dogs to control. Yet he was also potentially the best dog I had. He had become my first choice for many jobs, but only if I had the energy to keep on top of him. He was supremely strong-willed and never one to stand still for long. His sudden leap at me from behind was typical, an undiplomatic but highly effective reminder that perhaps we should be getting on with the job in hand.

The first element of that job was gathering the flock of a couple of hundred lambs that were grazing on the hill opposite. It had been two weeks since the blood-testing, and at last the results had come through. As expected, everything was clear and I'd been granted the licence giving permission for the lambs to be moved. I'd bring them in today to check their feet and run them through the footbath. The transfer could then be made on the day stipulated on the movement order, in less than a week.

The flock was separated from where we stood by the narrow shave of wood that made up the uppermost reach of this finger of the valley. Time was in short supply so my hope was that Ernie would be able to find his way through the trees, gather the flock, then drive the lambs so that they would be waiting for me at the gate to the woods.

'Look, Ernie,' I said to him quietly. As his eyes strained to the horizon he seemed to spot the distant sheep.

16

'Go through' was a command I used when the dogs were required to pass through a fence or gate. With a gentle 'Go through' from me, he was away, heading towards the woodland and then disappearing under the fence.

I put my whistle to my mouth and blew a long hard right-hand command, then waited for him to reappear in the field on the other side of the woods.

After a minute had passed I still hadn't spotted him so I blew again. When he still didn't come into view I began to scan the horizon in other directions. The behaviour of a small group of sheep in one corner of a field far to my left, raised my suspicions. They had lifted their heads and were looking intently towards the field edge beyond. A moment later the whole flock began to run together.

'Oh Ernie,' I muttered to myself, 'right idea, wrong location.' While I'd been looking elsewhere, Ernie had somehow managed to clear three stone walls and run across an empty field before arriving at a point perhaps eight hundred yards away. He had wasted no time getting down to work. The flock was already gathered and he was starting to use all his power and athleticism to bring the wrong sheep back in the most direct route to me.

'Ernie, that'll do,' I shouted across the valley. 'That'll do.' There was no response.

From a very young age, Ernie had possessed the most amazing natural ability to control and move a flock of sheep, although perhaps it would be more accurate to say that he was possessed by the instinct. Controlling him was always a battle of wills. Now, at this distance, he was enjoying himself immensely, and was oblivious to anything that I had to say.

I had to get through to him somehow so I blew a firm 'Stop' on the whistle. It worked and Ernie paused briefly, but the effect proved short-lived and he was soon succumbing to temptation once more. A few seconds later he had tucked back into his work and had soon driven the sheep across the field to the point closest to me where, luckily, they were restricted by the fence that ran between us. Content, he held them there as they moved to and fro.

'Ernie,' I growled. 'Listen.'

Eventually he did. Having burned off his pent-up energy, he responded to my calls and turned at last in the direction in which he had been sent ten or fifteen minutes earlier. Even from my position I could now see his tongue lolling heavily from his mouth as he made his way up the correct hill, but no sooner had he spotted the lambs than he was hitting top gear once more.

I made my way through the woods to the gate through which the lambs needed to pass. I had just opened it as Ernie arrived with the flock over the brow of the hill. Having achieved his objective, he lay down, his chest pumping heavily, great jaws open wide as he panted for breath.

'One day, Ernest, you'll make a great dog,' I said, 'but only if you bother to listen to me.'

As if to prove that day was still some distance away, he rose to his feet once more, and set off in pursuit of the flock that were now following the woodland path towards the farm.

Back at the farm, I penned the flock in the handling yards, then went to check on the lambs that had been suffering from pasteurella. In all, the outbreak had now claimed eight victims. Despite my best efforts, a further two had died in the pen in the barn and two more had fallen victim in the fields of Mortehoe. However, just as the storms had subsided, so too had the severity of the outbreak.

Here at Borough there were now six occupants in the pen. Of these, three were looking fit enough to turn back out into the field and two were tentatively eating. Only the largest lamb still looked hollow behind the rib cage. Inspecting him again, his stomach seemed to hang low, but his breathing was all right, and he remained stable, much as he had been for the past two days. He had a good chance of recovery. I felt a quiet confidence that the tide had turned, and things would continue to improve in the days ahead. Tomorrow I intended to give the entire flock their second vaccinations, hoping to deliver the disease a final, knockout blow.

I was just emerging from the shed when I heard the children arriving home from school sounding unusually excited.

Laura was the first to find me in the sheep-yards.

'Dad, Dad, there's sheep out on the road at the top of the drive,' she said, oblivious to the fact she was stomping through the mud in her shiny new school shoes.

My face dropped as I groaned to myself.

'Are they ours?'

'Mum says they aren't, but I think that they might be. You'd better go and have a look.'

The phone was already ringing by the time I reached the house in search of the Land Rover keys. A neighbour had spotted the errant sheep and she too suspected that they might be ours.

I was grateful for the call, but still had my doubts that the sheep were mine. Apart from anything else, since arriving at Borough Farm I had spent endless winter months erecting a boundary fence around the perimeter of the land. However, not everyone seemed to think that secure fencing was worth investing in, and there were still a couple of sheep farmers locally who were not perhaps as vigilant as they might be – in a couple of fields around the village sheep breakouts weren't uncommon.

Of course, while breakouts had always been a nuisance, the difference now was that with the movement restrictions still in place from the foot and mouth epidemic, the problem took on a whole new gravity. Reluctantly I jumped into the Land Rover and headed up the drive.

I'd barely turned out of our lane on to the main road before a dozen ewes loomed into view. I breathed a sigh of relief almost instantly. They were Scottish Black-face ewes, a breed usually kept on higher ground, and certainly not mine. In fact they were rather on the scraggy side, and I was mildly offended that anyone might have thought they were.

I pulled the truck up on to the grass verge. The traffic on the road wasn't heavy, but it was moving fast. I certainly wasn't going to risk the life of a sheepdog to retrieve this motley flock.

As I tried to decide on a course of action, one of the sheep lifted her head and, no longer satisfied with the plentiful roadside grass, trotted out on to the tarmac. There was a loud blast of a horn as a car swerved to avoid hitting the sheep. The noise completely spooked the rest of the small flock and they were off, scurrying down the middle of the road behind me.

Usually sheep that have broken out have a reasonable idea whence they came, so that, when they feel threatened, they will try to return as quickly as possible. On this occasion, however, they appeared to have little idea of where they belonged. Neither for that matter did I. The one thing I did know was that they couldn't be left where they were, so I reluctantly turned the Land Rover around in pursuit.

The late November evening was drawing in and the car head-lights were already shining. One by one drivers spotted the unexpected hazard, and slowed to a crawl. I put on my warning lights, just as the lead sheep made a bolt across the oncoming traffic and into a paddock alongside one of the few houses on that stretch of the road. A couple of passing motorists hurled a torrent of abuse in my direction. I smiled sweetly, cursed under my breath and followed the escapees into the field.

By chance the sheep had found a small paddock, unfenced, but rich in lush autumn grass. I parked the truck and tried not to spook them any further. Different breeds of sheep will behave in slightly different ways. The Scottish Black-face have a reputation for being determined runners. I could see a wild, unsettled look in their eyes; these were in the mood to take off at the slightest provocation.

In the mid-distance there was the sound of a slamming door, and the owner of the house appeared. He was an elderly gentleman, who I had seen frequently in passing but had barely spoken to before. I explained the situation, and in desperation asked if he had anywhere that we could pen the sheep. Needless to say he hadn't.

I was already regretting getting involved, not least because I couldn't now see how I was going to get out of it.

After ten minutes' deliberation, during which the sheep appeared to settle a little, I reluctantly accepted that I'd have to go home and fetch my sheep trailer, then take them back to Borough Farm. Someone was bound to claim them, I told myself. In the meantime I'd pen them in the sheep-yards, throw them some hay and wait. I left the rather bemused neighbour standing watch in the field.

It was almost dark by the time I got back to the ewes, but to my relief they were still grazing under the watchful eye of their newly employed shepherd. Parking the back of the trailer against a fence, I prepared for the next phase of the battle.

I'd brought the two youngest members of my sheepdog

21

team, Fern and Ernie. Such was my belief in the pair of them these days, I thought they'd be the best equipped to control a dozen determined Scotties. Working close at hand, Fern, like Ernie, had an exceptional talent for 'holding up' sheep. Between the two of them, I felt there was little chance of the fugitives escaping again.

No sooner had Fern and Ernie approached them than the flock were reacting. The lead sheep threw back its head and charged towards the far end of the paddock. Ernie couldn't hold himself back, and was soon gone to head her off.

With a 'Come bye' from me, Fern was then off to join battle with him, but by the time she reached the action, Ernie had already turned the head ewe and it was now stampeding back towards the trailer. Fern tucked the ewes on the left side in and forced the flock into a tight group. They were far from beaten, however. Each sheep turned repeatedly, trying to find a weak point in the dog's defences, but Fern and Ernie were more than equal to the task. What they lacked in their ability to be controlled by commands from me, they more than made up for in terms of their intense concentration and youthful speed and agility. After a few more turns, a couple more breakaways intercepted, and an ungainly collision with the fence, I slammed the trailer door shut with a sigh of relief. I was quietly pleased that the two youngsters had acquitted themselves so well. I thanked my neighbour and headed home with my consignment of unwelcome guests.

———

There was no moon in the sky as I walked back to the house that evening. What little cloud cover there had been had lifted, and an intense cold had descended, turning each breath into a strand of silver in the light shining across the yard from the house. It was eerily quiet.

The peace didn't last long. I kicked off my boots and climbed the dungeon stairs, to be confronted by a barrage of questions.

'Were they our sheep?'

'Did you get them all?'

'Can we keep them?'

My monosyllabic responses – 'no', 'yes' and 'no' respectively – dulled the children's interest almost instantly. Soon they were returning to more pressing matters, homework in the girls' case, and the sheep that had coincidentally escaped from his toy farm in Nick's case.

'Do you know whose they are?' Debbie asked, as I headed for the phone book and started a ring round of likely candidates. My answer to her was as succinct as the one I'd just delivered to the children. At this stage it was a definitive no. I didn't know anyone locally who had this particular breed of sheep.

Half an hour later, and having caused offence to a number of indignant neighbouring farmers, I was still no closer to finding out who the sheep belonged to.

'There's no mark on them or tag in their ears. I haven't got a clue where they came from,' I told Debbie, 'but someone's sure to miss them tomorrow, then they'll turn up to claim them.' I was trying to sound confident, but slowly the naive optimism of earlier in the evening was beginning to give way to a bout of worrying realism, and I could see just how many problems this situation was going to create.

The first thing that bothered me was that this might be construed as an official 'movement' of sheep on to the farm. In the wake of foot and mouth, the strict system of 'movement licences' was still in effect, and once sheep had been moved to a farm, no sheep could leave that farm for twenty-one days.

Since I intended to keep these interlopers isolated, and they weren't mine anyway, I rather hoped that this would not apply, but I already had a nagging suspicion it might.

There were other problems, potentially just as serious. The most fearful prospect of all, of course, was foot and mouth but the chances that these ewes were carrying it were now remote. All that having been said, however, there were still a number of highly infectious diseases that rogue elements like this could bring on to a farm. Given their parlous condition, there was a serious threat that they might be carrying one or more of them.

The condition I feared most was sheep scab, a voracious mite that eats mercilessly at a sheep's skin, causing it to rub relentlessly, or bite at its own flanks until they are red and raw. Unfortunately in its early stages the problem is all but invisible. Even keeping the ewes isolated from the rest of the flock may not be enough: the mites can apparently live on fencing stakes and other pieces of wood for up to three weeks, before finding a host. There is even a school of thought that the disease can be transported by birds.

As I spent the evening thinking about all the potential problems that these few sheep could cause, I felt like kicking myself for being so silly and getting involved.

In modern farming terms, Borough Farm and its one hundred and fifty acres is considered small, but in the not-so-distant past, it played a significant part in the local economy. The house and the rolling pastures that surround it were once part of a large estate, the property of the owners of Twitchen, a country house half a mile outside Mortehoe.

During its heyday in the nineteenth century, Twitchen was one of the area's grandest homes. Even though its owner, a Colonel Longstaffe, visited the house for just six weeks of the year and spent most of his time in London, a dozen staff were retained to look after it: eight as indoor staff and four as gardeners to tend to the extensive parkland. The gardeners were also responsible for the fruit in the heated greenhouses and peachery at the rear. The fresh produce was then dispatched by train up to the kitchens of the Longstaffe home in London.

Borough Farm was one of a group of farms tied to the estate that ran all the way from Mortehoe to the village of Lee, a few miles along the coast. The tenant would have kept a flock of no more than a hundred and fifty ewes, a dozen milking cows and twenty or thirty fattening cattle. That amount of stock is a fraction of what I need to make a living today, but such were the economics of farming back then, the tenant here would have been able to employ half a dozen men to look after the farm and its animals. Every once in a while I wonder what I might be able to achieve with that sort of man power.

The thought crossed my mind again next morning as, with the small trailer behind the Land Rover clattering noisily to the sound of bouncing shovels, spades, winches and clamps, I bumped across the steep banks opposite the farm. The errant ewes had been put to the back of my mind temporarily, and there were still several hundred yards of fence to erect before lambing in the spring.

A prominent feature of this part of North Devon are the Devon banks that divide most of the fields. Great barriers of earth, faced on either side with local slate, these monumental structures were built around three hundred years ago, partly using the forced labour of prisoners taken during the Napoleonic wars. Often they would be planted with a beech or ash hedge on top. As the banks are usually six foot wide, there is plenty of room for the roots to take. Indeed many of the hedge plants have since grown into massive trees that form a distinctive feature of the landscape.

During the Twitchen era, the farm would have had access to a small army of labourers to maintain the dozens of miles of Devon banking that ran around the farm's perimeter. Now it was down to just me.

At least there was some form of help at hand. More and more of the government money available for farmers is aimed at enhancing the environment, and to repair and protect historic features like this. I had been signed up to the Countryside Stewardship Scheme for several years, and had committed to restore over three thousand yards of banks, protecting each one with a sheep fence on either side.

It's a time-consuming job, but one which I rather enjoy. Although for some time after restoration the banks can look stark, the hedges grow back within a couple of years. I find it highly satisfying to play a part in the regeneration of the countryside.

I'd brought Greg and Swift with me for the day. For all the progress Gail, Fern and Ernie had made, my senior dogs remained the two whose company I most enjoyed, and while Greg ran alongside the Land Rover barking as playfully as an eight-year-old dog can, Swift trotted a little way behind, sniffing at the odd clump of grass en route.

I pulled up by a large bank topped with a sprawling hedgerow, defined on the Ordnance Survey map as the parish boundary. For some reason this boundary hadn't been hedged by beech or ash but instead with goat willow. This had thrived, and now spread untidily, overhanging the field by many yards on either side. The bank had probably once contained sheep and cows rather efficiently, but the expansion of the roots – not to mention the heavy rain over the years – had taken its toll. On top of that, sheep that for many years had been able to climb through its rather weak defences had eroded the bank. It only added to its unkempt appearance. All in all it was in a sorry state.

The first job was to cut back the overgrown foliage drastically with a chainsaw. Unfortunately the middle of the bank resembled a scrap-metal yard, and was filled with pieces of corrugated tin, barbed wire, and just about every other conceivable object. Someone had even put an old bedhead in here to block the gaps at one point. Such pieces of iron are the chainsaw's worst enemy, so it was only after I'd spent half an hour removing the detritus that the saw finally roared into action. It still wasn't easy going, mainly because of the amount of thick bramble shoots present, but after an hour I'd made good progress.

The unusually wet autumn had left the stems of the brambles heavy with ripe blackberries. Turning the saw off for a minute I helped myself to a handful of the fruit left unpicked by the Valley's blackbirds and finches. Greg immediately spotted me and was soon idling over, his great bushy tail wagging in anticipation. A few yards away from me, however, he stopped in his tracks and began sniffing at the base of the bank under the brambles.

I walked over to see what had attracted his attention. The rancid smell emanating from the spot soon told me it wasn't anything pleasant. The stench was being caused by a decaying fox that lay among the leaves. It wasn't difficult to see the reason for its demise – mange. Much of its body was bald and raw, even the backs of its ears bore the signs of claw marks from repeated intense scratching. Mange was endemic in the local population and I'd now seen half a dozen victims this year. Mother Nature can be harsh, and perhaps this is just her way of controlling a fox population that had become a burden.

There's always a chance of fox-mange transferring to dogs,

so I kept Greg and a suddenly curious Swift at a distance while I cast some earth over another victim of this blighted year for the countryside.

———

By mid afternoon the lingering warmth of the distant sun was already being replaced by a creeping cold. In the fading light of the valley below, a blackbird clucked in alarm, as if complaining about the drop in temperature. Gathering up my tools I made my way back to the farm, passing the pen holding the new arrivals in the yard as I went. 'They're still here then,' I muttered to myself as we headed past. 'I suppose I'd better have another go at finding where they belong.'

Almost twenty-four hours since they'd arrived on the farm, the stray ewes had, if anything, become even more of a mystery. I'd exhausted my list of farmers to call and was already beginning to resign myself to an unwelcome, prolonged period of ownership. For a moment, however, I put myself in their owner's shoes, and tried to think of whom he might contact if he were trying to reclaim them. I hadn't yet called the police and this now seemed the common-sense thing to do. They were often people's first port of call when they lost animals. Maybe they'd sort it out, I told myself.

To judge by the disinterested tone he adopted on the phone, the prospect of helping out 'Little Bo Peep' obviously didn't fill the duty officer with much enthusiasm. In truth, the local police station had probably had its fill of farm problems for the year. Throughout the spring they were diverted from their normal work to oversee the application of the foot and mouth regulations. Instead of looking for criminals, the local constabulary had had to supervise animals crossing the road, cordon off infected farms and deal with people who chose to ignore the closure of footpaths.

'No one has reported any lost sheep to us I'm afraid,' he sighed, after what he claimed was an exhaustive search of his records. 'I'll take your name and address, and give you a call if we hear anything. You could try the RSPCA,' he said helpfully. 'They'll sometimes re-home stray animals.'

I'd thought of the RSPCA and had almost dismissed them as a potential solution already. These sheep were lost rather than injured, so weren't really likely to be of their concern, but it was worth a try, I guessed.

The response was just as I'd predicted.

'Are they distressed or injured?' asked a very concerned lady on the other end of the phone.

'Well not really, but they're not in their first youth either,' I said, clinging to the hope that elderly animals fell under their auspices as well.

It was a vain attempt.

'I'm sorry but we're so busy at the moment, unless there's a problem with the animals, we really can't help.' She was obviously well intentioned, however. She thought for a moment. 'I could send an inspector down to see if they are injured or distressed.' She was trying to be helpful, but I really didn't feel the need for a second opinion.

'No, don't bother about that. I think I would have spotted if anything was wrong with them.' The despondency in my voice was obviously beginning to show.

'I tell you who might be able to help you,' she said brightly. 'The Ministry at Exeter.'

Reluctantly I put the phone down, then picked it up again to call the Ministry, newly renamed DEFRA.

After being passed through half a dozen phone extensions, I eventually reached someone who could help – or at least thought that they could. Wearily I explained the whole saga once more.

'All I really want is to get them off the farm, but without knowing whose they are I can't get rid of them,' I said.

'I see,' said the voice, exuding a lack of any real interest.

But then I made my fundamental mistake.

'Is there a temporary ear tag I could put in them, so that they can be authorised to go off the farm?'

The moment I said it I realised I was heading down another, more perilous route. The voice on the other end of the phone suddenly became imperious.

'Mr Kennard, as you should be well aware, all sheep must be tagged with *your own* flock number before they leave the holding. It is an offence to do otherwise.' He thought for a moment. 'I'll have to inform Trading Standards that you have the sheep on your farm.'

'Thanks,' I said rather grumpily as I hung up.

It wasn't long before the phone was ringing.

'Mr Kennard? I understand that you have some stray sheep on your farm,' announced a voice introducing itself as an official at the Trading Standards Office.

For a moment I thought I was getting somewhere.

'Ah yes, can you help in finding out where they came from?' I asked.

'Have you got their ear-tag numbers?'

'No, they didn't have ear tags on them.'

'Well, you shouldn't have moved them on to the farm without an ear tag, you know. It's a serious offence,' he said.

This was turning surreal, I thought to myself. I couldn't bite my tongue.

'But I only moved them on to the farm to prevent them causing a serious road accident.'

There followed a lengthy discussion in which I described the precise details of what had happened. Throughout, the voice on the other end of the line remained unmoved. By now

I was wishing I had just left the sheep in the middle of the road, but eventually the penny seemed to drop and the voice on the other end of the phone relented.

'Oh, I see. Well, we'll overlook it this time,' he said. My relief was short-lived, however. 'But I will have to put your farm on a twenty-one day standstill,' he added, as if he were doing me some huge favour.

I protested for a few moments, but I knew that anything that I said was likely to be in vain. I gave up and instead set about trying to solve the issue of what to do with the stray sheep now.

'Isn't there any way you can help me get them off the farm?' I asked in exasperation.

'It's not our area to deal with.' Then, as if struck by inspiration, he added, 'But I tell you who might be able to help. Why don't you give the police a call?'

As I made myself a calming cup of tea, I realised this was a bad situation that was getting worse by the minute. Since passing the blood test, the lambs I wanted to move had been issued with a movement licence. It now dawned on me that the twenty-one-day standstill order the Trading Standards officer had just slapped on would preclude me from making this move. I knew there was a time limit on the licence, and this might run out before I could move the sheep.

I headed for the office. Thankfully the system had been revised a little recently: during the early part of the year, licences had run to seventeen pages of fax paper; now, at least it was down to two or three. I ran my eye over them and soon had my worst fear borne out: the licence not only expired long before the end of the three-week standstill, it also specified a date of movement which fell within the prohibited period. I couldn't move the lambs on any other day.

If I'd had a waste-paper bin I'd have kicked it.

Thanks to these wretched stray sheep, my farm was once

more going to be reduced to a state of semi-paralysis. Nothing was going to be able to move off it without a new blood test certificate. We'd have to go through the whole testing rigmarole again – just when I thought we might finally be turning the corner. I'd spent the day working on a barrier to stop my flock breaking out, but now it felt like we were all fenced in.

CHAPTER THREE
'That Dog'

For a livestock farmer, Saturdays and Sundays are little different from the other five days of the week. Animals still need to be checked, fed and watered, and any problems need to be dealt with.

During school terms the main difference for me is that the children are on the farm at the weekend. Their presence had always been a pleasant distraction to the repetition of the daily routine, but during the dark days of the past year they had provided a breath of fresh air.

Today, as the Saturday morning breakfast was hurriedly

cleared away, there was a hazy glow in the morning sky to the east that suggested the enveloping mists of the last few days might finally be lifting. The children had already spotted this and were getting ready to take advantage of the break in the weather.

My eldest daughter Clare had already pulled on her boots and was headed off towards the stables. After many years of nagging, Debbie and I had found her a pony. She'd served her apprenticeship by taking riding lessons in the village each week for more than four years and seemed ready for the next step, then, a year ago, we'd been offered a rather elderly but sprightly pony on loan. Minty seemed to fit the bill perfectly for Clare. He was easy to handle, steady and well behaved. The only down side to his character was that, as his previous owner had told us, he 'hated men'.

With Clare already out of the door, Nick was the next to appear. Now five, over the last year he had become more and more enthusiastic about his days helping me. 'Dad, what are you doing?' he asked, as if he suspected that I might be thinking of going to work without him.

'Going round the sheep out on Morte Point,' I said, pausing for a moment. 'Do you want to come?'

I didn't for one minute think that he'd say no, and instantly a great smile spread across his face.

'Yes,' he said, jumping up and down on the spot. Then, just as suddenly, his face dropped, and he looked serious for a moment. 'Dad, will I need my long-job wellies or my short-job wellies?'

For some reason known only to himself, Nick had recently adopted a two-pair system of Wellington boots. Debbie and I were at a loss to understand the difference, but to Nick it was of vital importance. I thought for moment, before I replied in a serious voice.

'Oh, long-job wellies this morning I should think, Nick,' I told him as I headed out into the yard, 'and don't forget to tell Mum that you are coming with me.'

As I crossed the yard, Minty was already greeting Clare from the stable with an impatient but restrained whinny. In comparison the morning chorus of my dogs was rather more vocal. I opened the doors of the run and as usual five excited barking dogs sped off up the drive, chasing one another wildly. Then they returned a few seconds later, and lined up expectantly, side by side in strict order, still barking, with only Greg moving forward towards me.

I never ceased to be amazed at the way in which four of the five dogs in my small pack established and maintained their tightly run hierarchy. Despite his advancing years, Greg was still top dog and the other three arranged themselves accordingly. Alongside him, Swift lay crouched on the ground, her eyes glued to her pack superior's every move. Next to her stood Fern, waiting for a movement from her adopted matriarch, then came Ernie, the rookie, looking at his immediate superior.

Only Gail stood slightly disconnected from this quartet, as ever making the point that she refused to fit into this pecking order.

Gail harboured a deep distrust of Fern. They were of a similar age, and had never established a superiority between themselves. Every now and then a minor dispute over a bone, or some other seemingly trivial matter, would spark a serious row. It would rarely happen if I was around, but I was careful never to leave them alone together, even in the back of the Land Rover. Today, as always, as I opened its back door, the dogs split into two groups, Fern and Ernie ready to come in on one side of the divider and Greg, Swift and Gail on the other.

I'd separate them in this way for good reason. Apart from

Gail and Fern, the only other tension was between Greg and Ernie: the elder statesman clearly saw the youngster as a potential rival and by far the biggest threat to his previously unchallenged position.

It hadn't happened this morning, but the tension often showed itself during this first interaction of the day. Ernie might start to walk in my direction, looking for a little attention from the boss. Greg would see this and immediately make a darting run across his bow, raising his tail and hackles in an aggressive pose. Although Greg was now well past his prime and Ernie was at the peak of his fitness, the young tyro was still unwilling to challenge the top dog: he would cower, turning his head away in apparent submission. How much longer this would continue, however, remained to be seen.

I was just about to slam the door of the Land Rover on the five dogs, when the dungeon door burst open.

'Don't go without me, Dad,' Nick shouted, a broad grin on his face, and was about to clamber up into the passenger seat when we heard Debbie calling from the house.

'Nick, you haven't got a coat,' she shouted, 'and how long are you going to be?'

Nick looked at me and then down at his feet with youthful exasperation.

'Mum! Can't you see? I've got my long-job wellies on,' he said.

A spectacular landscape of steep-sided valleys and wooded combes that meet the sea amidst gorse-covered heathland, the coastline around Mortehoe is one of the most beautiful in the country. I count myself lucky to work against such a spectacular backdrop. Much of it is owned by my landlords, the National Trust, including the eight-mile section of the

Morte estate, which stretches either side of Mortehoe. My tenancy with the Trust includes the two hundred acres of Morte Point, a peninsula reached by a narrow lane, that runs from the square-towered church in the middle of the village, past the churchyard.

The Point itself is accessed by a pair of large iron gates, set on square granite posts. I pulled up the Land Rover and asked Nick to jump out and open them. One of the gateposts bears an inscription to Lady Chichester, who donated Morte Point to the Trust in the 1920s, but as he struggled to haul them open, Nick wasn't too bothered by their history.

'I can't do it, Dad,' he groaned, leaning almost horizontally back as he put every ounce of effort he could into shifting the obstacles.

'They're a bit heavy for you yet, I reckon,' I said, stepping in and swinging the gates open with their familiar squeak. Then I lifted him back up to his seat.

I stopped the Land Rover a few yards on, ready to start the morning's inspection on foot.

'What do you think, Nick? Ernie could help us this morning, couldn't he?' I called to my assistant as he wrestled open the door, turned on to his belly and slid down from the Land Rover again. He looked at me with a mixture of excitement and disbelief. Along with his sisters, he'd listened with glee to the story of Ernie rounding up the wrong field of sheep a while back.

'Dad! I don't think Ernie is a good enough sheepdog,' he said.

For a moment, I wondered whether his judgment might be better than mine on this occasion. With over two miles of unfenced cliffs, Morte Point is no place for the inexperienced dog. Although in many places these cliffs are no more than a scramble to the sea, there are numerous gulleys and inlets, where the drop can run to two hundred feet. Yet I had a strong feeling that I now had enough control of Ernie to be able to work here. Besides which, I had to think of the future. I wouldn't always be able to rely on Swift, Greg or even Gail and, while it's always easier to take the most experienced dogs, there comes a time when the youngsters must be given a chance to prove themselves. So, as I opened the door to his side of the Land Rover, I called, 'Just Ernie.' He leapt out eagerly as I forced the door shut before the others could join him.

The morning was cold and the breeze coming in from the sea to the south-west wasn't strong enough to blow away the last of the fog that still hung over the highest outcrop, shrouding the site of the old coastguard lookout. The gorse dripped great shining droplets that clung to every sprig. Spiders' webs, spun over the sharp spikes, sagged under the weight of the morning dampness.

Today's job was just a part of the usual routine, checking the stock. Though nineteen times out of twenty, this reveals nothing of consequence, it's something that simply has to be done. Sod's Law will ensure that the one day you skip it is the day that some problem or other occurs. I turned up my collar and pulled Nick's woolly hat down over his ears. We headed off towards the path that would take us along the south side, where the bulk of the flock habitually graze.

The stiff breeze numbed the ears as we walked over the brow, looking for a suitable vantage point from which to view the flock spread out below. Nick stopped to tighten his coat around his neck, and pressed a cold hand into mine, but he didn't complain.

The bracken was never as tall on the south side of the Point as on the north, and at this time of the year its dark amber stems were broken down to no more than boot height. The flock were clearly visible and for a moment I thought about sending Ernie to turn them in so I could get a closer look, but I quickly decided it was a tricky run for him to take and resorted instead to casting my eye across the contented sheep for a few minutes, until I was satisfied all was well.

The bracken may have been in its annual retreat, but the gorse was as bountiful as ever. Each year more and more of the peninsula was being engulfed by its dense sprawling clumps. Traditionally gorse is kept at bay by 'swailing' or burning a small area each year, but this has become unpopular with modern conservationists. The problem is that other methods are all but impractical on such a steep and rocky area. Hence its seemingly relentless encroachment on to the Point.

The great areas of gorse weren't to everyone's liking, but, judging by the tac-tac-tac of a stonechat coming from within the overgrowth, someone was content there at least. Nick just managed to pick it out as it sat noisily perched on an upper sprig. We watched him for a minute until he dived into the depths of the overgrowth, retreating to his own world, hidden from view and away from predatory birds.

The tip of Morte Point reaches deep into the Atlantic, and is dominated by a rock slab spine, that forms a north and south facing slope on either side. The constant windburn and sea spray mean that the pickings are sparse for foraging sheep, but for some reason it is an area that they graze constantly. As always a dozen or more ewes were among the rocks at the top of the heather-strewn slopes.

'What do you think, Nick? Do you reckon Ernie could round those sheep up for us?'

Nick looked amazed at the prospect of my least controllable sheepdog being even considered for such a thing.

'Daddy, he might get lost,' he eventually replied after some thought.

Ernie, on the other hand, appeared to have no such doubts about his own abilities. He stood poised and alert, staring intently at the sheep.

'I don't think he'll get lost, Nick, and at least he can't go around the wrong field,' I said. The path that he would naturally follow was a long way from the cliff edge and would bring him up behind the sheep, so with an 'Away, Ernie', I set him off.

Ernie raced away with all the caution of a circuit greyhound chasing a hare. The arc he'd chosen was rather wider than I would have liked. Instead of starting off along the path in front of where we stood, he dived to the left, and started picking a route along the cliff edge itself, heading down on to the rocky ledge where the soil had dropped into the sea.

He had only gone a couple of hundred yards when he climbed from his low level path, and pulled up. He had clearly spotted the sheep, now some way above him, and was unsure of what to do. I put the whistle to my mouth and blew a long hard left-hand command to send him further along the cliff. He heard the whistle and was on his way in a second.

Sometimes fate just seems to conspire against you, and unfortunately Ernie wasn't the only one who had heard my shrill blast. A ewe that had been grazing lower on the slopes than the rest lifted her head and pricked her ears in the same instant. Sensing an approaching dog, and finding herself isolated from the rest of the flock, she panicked and turned to run flat out along the path, in exactly the direction of the end of the peninsula. Her break for freedom sent her beyond the point where Ernie was heading ready to sweep in behind the flock. Crossing the path, he found himself twenty yards behind her. This was more than he could bear. Every instinct in his body told him that he must head off the errant sheep, but I knew I risked a serious mishap if he did so.

I abandoned the whistle for a second, and tried using the impact of a loud, bellowed: 'Stop you!'

He paused just for a second. The section of the path that the ewe now found itself upon was narrowing as it got closer to the cliff edge, three hundred yards from where we stood, and she was soon out of sight. For Ernie losing sight of her was the final straw and, ignoring my repeated commands, he was off in pursuit.

Nick had been watching the events with a growing sense of alarm.

'Dad, what are you going to do now? I told you Ernie wasn't a very good sheepdog.' I didn't bother to explain.

'Come on, Nick, we'd better go and get him,' I said instead.

Somehow the run along the path seemed to take for ever. Although Nick trotted along well, he couldn't help but hold me back. My panic was mounting.

Hours seemed to have passed before we reached the narrow path. Further ahead, at the very tip of the peninsula, it turned acutely back inland, with only the rocks and Atlantic beyond. The path here was no more than thirty feet above the sea, and at the tip the rocks sloped gently down to the water's edge. The sea was quite gentle. Lazy waves cast spray a few feet on to the rocks as I scanned them, fearing that, in her attempts to outpace Ernie, the ewe might have run straight on into the sea. I satisfied myself there was nothing to the left, but before I could look to the right, Nick had spotted something.

'Dad, look,' he called, his voice a mixture of excitement and concern.

I followed the direction of his pointing finger and there down below us was the ewe. She had chosen to effect her escape by running along a line of rocks that formed a sort of causeway from the cliff path down to the sea. There she now

stood as sheep often do when feeling the effects of exhaustion
– stationary, her gaze directed towards the sea.

'Look at Ernie!' cried Nick.

Of far more concern to me than the ewe was the precarious
situation in which Ernie now found himself. So strong had his
instinct been to turn the ewe back towards me that, when she
had taken herself down to the water's edge, he had jumped
perhaps six feet across a narrow gully that lay directly behind
the ewe. Now he stood balancing uneasily on a sharp ridge of
rock, only a few feet above the water which surrounded him on
all sides.

The narrowness of the gully meant that, with the ebb and
flow of the water, each new wave threatened to wash him from
his perch. Despite this Ernie stood still, transfixed by the ewe,
which he was still determined he must prevent from escaping.

'That dog!' was about all I could bring myself to say to Nick,
as he looked on with incredulity. Obviously there was an
expletive missing from that sentence.

Together we scrambled down in the direction of the ewe.
Then, leaving Nick to sit on a convenient rock, I positioned
myself between her and the sea. Sensing the game was up, she
turned and picked her way back up towards the path above.
Without the sheep to focus on, Ernie had turned his attention
to his own plight, and obviously didn't much like it. The
occasional wave was breaking over one end of his tiny island,
wetting him with spray. He now began to move awkwardly,
looking for a foothold from which he could jump across. His
claws were finding little grip on the wet rocks, and with each
movement he threatened to slip.

I weighed up the options. The gully was one of those that
was made up of the deep gap between two steeply sloping
rocks. I could see the bottom, but it was perhaps three feet
deep, so the thought of wading in to get him didn't appeal.

Although the rocks on the land had supplied enough grip and surface for Ernie to spring across, there was no way he could now jump back.

'He'll have to swim across,' I called to Nick, who had started picking his way over to join me.

'But Daddy, can Ernie swim?' he asked wide-eyed, a massive grin spreading over his face in anticipation of the entertainment to come.

'All dogs can swim,' I reassured him, as I positioned myself at the narrowest crossing point, and started to call, 'Come on, Ernie, you got over there; now you'll have to swim back.'

Ernie swayed a little, trying to pluck up the courage to take a plunge, but each time he appeared ready to go, a wave passed through, sending him scrabbling backwards.

'What happens if he can't get back?'

Nick's 'what if' questions were something of a speciality at the moment, and usually the result of an imaginary set of circumstances.

Before I could answer, Ernie had inched forward once more, but on trying to retreat this time his claws were unable to grip. He slid unceremoniously into the gully, and began swimming frantically across.

As he did so another wave carried him along until he rose almost to within my reach. Then the level fell, before rising once more. This time I could just about get my hand on him, and, grabbing at the scruff of his neck, I lifted the sodden dog from the water. He clawed at the rock and began to find his foothold, but, as I relinquished my hold on his neck, somehow my balance was gone. Trying not to fall over backwards, I spun around, but found nowhere to grip. A second later I stood waist high in the freezing November sea.

Instinct took over. The breathtaking cold propelled me towards a low point in the rock and I began hoisting myself up. I looked towards Nick and found it had almost been too much fun for him to take. Bent almost double, he had tears of laughter rolling down his face.

I pulled off a Wellington boot and emptied the contents. Nick at last found some composure.

'Daddy, why did you do that?' he asked, before erupting into another fit of giggles.

Ernie on the other hand was less impressed. The soaking I had taken on his behalf was apparently of no interest; his gaze had once more returned to the ewe. Having reached the coast path, she looked down at us for a second. Quite satisfied with her morning's work she trotted off in the direction from which she had come.

Ernie turned back to me for a second, and I could tell that he was contemplating another attempt to turn the ewe.

'Ernie, no!' I growled at him, as I emptied the water from the other boot. This time he thought better of it and stopped in his tracks.

'Have we got to check any other sheep now?' Nick asked, taking my hand once more as we made our way back up the rocks, my boots emitting a rhythmic squelch as we walked.

'I think that the rest will be all right for today,' I said.

Nick thought for a moment.

'But what if a sheep . . .'

As Nick's question drifted off into a long imaginary scenario which involved a boat, a dog, and a very long swim for me, I focused my attention on the distant, off-white speck of the Land Rover, which at that particular moment seemed a very, very long way away.

CHAPTER FOUR
Mixed Fortunes

Every time I looked at the dozen stray ewes, still standing in the pen in the yard, I couldn't help but feel the resentment welling up inside me. They'd been with me now for well over a week, and today, as I tossed them their afternoon rations, I muttered away to myself about the problems they'd caused in that time. Needless to say, my grumbling had no effect whatsoever on the sheep. They just munched away noisily, their teeth grinding, oblivious to the world around them.

Fern pushed her head through the gate, eyeing them warily,

and the old ewe nearest to her stirred herself to stamp her foot as she chewed on a mouthful of my best hay.

'Here, Fern,' I said, calling her back then walking away towards the lower fields. She looked at the sheep, considered her options for a moment then scurried across the yard behind me.

She was now over four years old, but had yet to fulfil her early promise. Her training was virtually complete, and she had developed into a fine dog for work both at home and potentially on the trials field – but only when she felt like it! Though to be fair to her, she'd had no chance to prove herself in competitions over the previous summer. Along with just about every other rural activity, sheepdog trialling had been suspended for the year.

There were rumours that it might return for the next season, beginning in the spring. If it did, I was determined to run Fern in as many trials as possible. In the meantime, a little extra training here and there could make a lot of difference. There had always been a curious lack of confidence about Fern, but if she was sure of the job that she was doing, then she was fine. She would run like the wind, take every command and work with intelligence.

These strengths had certainly shown through in her few competitive outings before the sport was put on hold. Sheep are not used to being worked in small groups, and when split into groups of five for a sheepdog trial they often become flighty. Fern had proved herself an expert at working carefully without 'spooking' the animals. In particular her wide turns when behind them, made without cutting corners, were near perfect.

Yet she also had a real weakness. Nine times out of ten, once she had arrived behind the sheep, she would put on a good show. The problem was, Fern had to get to the sheep in the first place, and that's where her main problem lay.

When setting a dog off from the start – or 'post' – a handler tries to give an indication of the direction in which it is being sent. Far too often this was wasted on Fern. She never seemed to be sure of where she was going, and would frequently set off with a look of bewilderment about her. It was very frustrating, and not something that I could easily remedy, but whenever there was a chance I took her for a little extra training on her own, concentrating on building her confidence by giving her tasks that she was certain about.

I really enjoy training a sheepdog. Gently moulding the natural ability that already lies within the dog is an art, and it's a hugely satisfying feeling when a dog is learning fast, and the two of you are beginning to work together as a team. This evening, however, it also felt therapeutic. Working with Fern was going to give me a few minutes to forget about all the hassles of the past weeks.

I watched her with satisfaction as, for once, she worked the way she was capable of doing, sweeping off along the bank of the old railway line far away to my right. I gave her a gentle whistle to slow her as she brought the sheep into view, and she tucked her head lower in concentration as she purposefully marched them forward.

A shrill blast of the whistle and she darted arrow-like to the right.

A long blast for the stop and she was down on her belly, itching to be invited to move again.

A higher tone this time, and she was away to her left, heading the ewes as they tried to break around the corner of a stone wall. She was there in a flash. This was Fern at her best, and it was a joy for me to see those years of training coming together. If only she could work like this in competition. If the trials were to start again next year, surely she would take some beating.

With the sheep now close at hand, I asked her to 'shed' them into two smaller packets. Shedding was one of her specialities, and she came through the crowd with such speed that the ewes split off into two groups immediately with one group bolting to the far corner of the field. This presented an opportunity to practise the most difficult command for a sheepdog to take. I called her on to the half of the flock that were still in front of me, and carried on working with her, giving her a succession of commands: left, right, out, in, stop, a yard to the left, two yards to the right. All the time I was watching the ten sheep that had now wandered to the top of the field.

I sensed the time was right and blew a low rising tone on the whistle – Fern's 'Look back' command.

'Look back' is the order given to a sheepdog, asking it to leave the sheep that it is working, and go in search of some hitherto unseen animals. It's considered the most difficult of all commands to master, mainly because a sheepdog tends to focus obsessively on the sheep that it is already working.

At first the signs were good. Fern spun around from the sheep, until she was facing back up the field, looking alert and intent. I followed up with a loud blast that should have sent her running to her left, back up the field to collect the other ten sheep. But she didn't budge.

Instead, she appeared to contemplate the idea for a moment, before turning back to the sheep in front of me.

My heart sank. She knew exactly what she was meant to do, but somehow just didn't think she would bother today. I tried again. The same low whistle, but this time she had made up her mind. With an air of finality she lay down, looked away and tucked her ears back.

This was a complete refusal. I knew that it was pointless carrying on with the command; Fern had decided that she wouldn't 'look back' and that was that. The fact I knew she was an expert at this manoeuvre on occasions only compounded my frustration. Today's training run had summed Fern up. In the space of two minutes, she had demonstrated the best and the worst of her abilities. I tried not to let my annoyance show through. There's an old adage: 'Never finish a training session on a bad note', so I ended with a routine that would restore Fern's confidence a little. She was happy to carry on with the sheep in front of her, and I let her work those for a few moments, before calling her back towards the gate.

Walking from the field, I couldn't help thinking I might have to temper my hopes for the forthcoming sheepdog trials season: 'She'll either take some stopping, or she'll take some starting, and there won't be much in between.'

Two weeks on since the lambs at Mortehoe had been vaccinated, and the pasteurella scare seemed to be over. Now at last I felt confident enough to pen the lambs, in order to do some of the jobs that I had had to put off during the outbreak. Many of the lambs needed 'docking' to remove the soiled wool around their tails, and all needed a worm dose, but the most significant job of the day – indeed one of the most important of the year – was picking out the ewe-lambs that might make suitable flock replacements.

Old age, loss of teeth and mastitis all take their toll on a flock, so it's usual to replace about 20 per cent of the breeding ewes each year. There are two options for covering the shortfall: either you 'run on' enough ewe-lambs to make up the numbers, or you buy in 'shearlings' in the autumn sheep sales. Since my arrival at Borough Farm I'd done a bit of both, but concentrating on building up a flock of North Country Mules. As their name suggests, the Mule is a hybrid ewe produced in the north of England, specifically by crossing the hardy Swaledale hill ewe, with the more prolific Blue-faced Leicester ram.

Over the last forty years these Mules have become highly sought after, mainly for their ability to deliver and rear a good number of lambs. Tens of thousands of them make the journey from northern markets to lowland farms each year. The only downside to the Mule has been their price, and their habit of prematurely losing their teeth, which in turn leads to a loss of condition. In the past I had considered this 'high input–high output' sort of sheep provided our best option at Borough Farm, but in recent years I had begun to question this approach.

Some of the traditional British breeds of sheep produce far fewer lambs than the Mule, but have the advantage of being hardier and lasting substantially longer. I had introduced one such breed, the Romney, to the National Trust land at Mortehoe and never regretted it.

The Romneys' general ability to look after themselves with a minimum of shepherding was very appealing, particularly given the way farming was going. It had been obvious for years that fewer and fewer youngsters were coming into the farming industry, but foot and mouth had accelerated the decline. Even before the outbreak, few people wanted to spend their life in mud, wind and rain, tending to some of the world's dimmest animals. Then seeing thousands of those animals destroyed, and the resulting chaos within the industry, had made it an even less appealing prospect.

In the future, I could see that farmers like myself were unlikely to be able to depend on extra assistance when we needed it. It only added to the attraction of sheep that lived outside throughout the year, required less feeding and survived longer, especially if they were cheaper to buy as well.

Until now I had kept the pure Mule flock at Borough Farm with a focus on producing the most lambs possible. However, with the cost and ease of maintaining a sheep now as important as its productivity, I had decided to change this approach and – for the first time at Borough Farm itself – replace the outgoing Mules with some Romneys.

I spent a couple of hours sorting out strong Romney lambs with a good-sized frame and a 'clean' wool-free head, identifying all eighty of them with a blue mark on the shoulder. Then I moved them to a separate paddock, before heading homeward. As I shut the gate, I ran my eye over them one last time. The aim today had been to pick a consistent-looking group of sheep. To me they looked almost like peas in a pod: a worthy addition to next year's flock at Borough Farm.

By the time I got back from Mortehoe, the evening was fast closing in and the sound of gathering rooks echoed up from the valley below. As I clambered out of the Land Rover, Debbie was waving at me from the kitchen window.

'Have a look,' she shouted, pointing to the sheds in the far corner of the yard. 'The stray sheep have gone.'

'What do you mean? They've broken out again?' I called in dismay.

'No, they've gone!' she said with glee. 'I was just getting home from school with the children and there was a van with a trailer on, pulling out at the end of the drive. I couldn't see who was driving.'

It had taken its time, but obviously the farming grapevine had eventually done its job. I'd spoken to so many farmers in the area, word had somehow spread to the sheep's owner.

'What sort of van was it?' I asked.

'An old one, on its last legs. Grey.'

From Debbie's description of the van and what had

happened, I already had a sneaking suspicion I might finally know who the mysterious owner was.

'If it's who I think it is, he probably waited until he thought the farm was empty then sneaked in and out without us seeing him,' I said.

Any sympathy I'd felt for the owner had long since waned. At the back of my mind I'd decided that if no one had claimed them when the three-week standstill was up, I'd ear-tag the ewes then sell them to one of the outlets I knew that was buying sheep. If someone had turned up after that, I'd have given him a cheque – less some money for hay.

Now, at least, I'd be spared any further hassle. The problems they'd caused had yet to be resolved, of course – I wouldn't be able to move the sheep that I'd agreed to sell until the twenty-one-day standstill order was lifted the following week – but it was a huge relief to be rid of them. As we headed back to the house, I couldn't help but smile at Debbie. 'I think I can safely say that's good riddance.'

The disappearance of the strays seemed to signal a small change in luck. The following morning I discovered a second encouraging development – in the most unlikely location.

Since the outbreak of foot and mouth, visits to the internet – and the DEFRA website in particular – had become a matter of routine. At the height of the epidemic it was a necessity – the rules and regulations affecting farming changed on an almost daily basis – but it hadn't been a place I'd associated with good news. On many an occasion, it was here that I'd learned of a new farm in the area having been hit by the disease. The announcements were made in cold, business-like language that could never reflect the personal tragedy that lay behind them. There were times when I had watched the home page flash up through half-closed eyes, fearful of what I might see. As I scanned it this evening, however, my reaction was altogether different.

Immediately a newly posted headline jumped out at me: BLOOD TESTS NO LONGER NECESSARY FOR SHORT-RANGE STOCK MOVEMENTS. As Julia, the Ministry vet had predicted during her last visit, DEFRA had decided to relax the rules on compulsory blood-testing before all movements of livestock. I read on, and was delighted to discover that this was going to come into effect within a week or so.

This was welcome news for all sorts of reasons. I was still restricted by the twenty-one-day movement order, but it was the need to blood-test before sheep could be moved from farm to farm that had been the biggest problem.

I'd agreed to sell a couple of hundred ewe-lambs to a friend, Martin, who farmed a few miles from Barnstaple. He had lost a thousand sheep and nearly two hundred head of cattle during the foot and mouth, and when we had last spoken he had talked about restocking his farm. Now that the blood-testing restriction had finally been lifted, I called him again to agree a price and to arrange when the lorry could come to pick them up.

Two hundred lambs sold off the farm. That at least was some positive news.

In truth, I had mixed feelings about the sale. On the one hand, I was delighted to be helping Martin to restock. He was a good friend who had suffered enormously during the past six months. It was a measure of the lengths to which affected farms had to go in order to get the all clear for restocking that he and two other members of the family had spent four months, power washing and disinfecting every building on the farm from roof to floor. Having started in May they had finished at the end of September. I couldn't even begin to imagine how important a moment it must have been for him to be ready to welcome livestock on to his land once more.

On the other hand, the price we'd agreed for the 200 lambs was far less than I would have hoped for.

This was no reflection on Martin. Lamb prices have always fluctuated wildly. In the mid 1990s they peaked to such an extent that farmers began to feel they might at last catch up with inflation and turn a profit, but the increase was short lived, and was followed by a slump that saw the value of a lamb fall to a level that hadn't been seen since the 1960s. It had stayed that way for the rest of the decade, leaving many sheep farmers feeling that there was no future in the business.

Ironically, in the six months before the foot and mouth outbreak, the price had rallied once more, but the epidemic had – obviously – sent the price into a violent, downward spiral. Even now, months after the disease was eradicated, prices were still on the floor. Now, the limited amount of stock going on to the market should have signalled a rise in prices, but for some reason the normal laws of supply and demand hadn't applied themselves. There hadn't been anything like the sort of surge in prices you might have expected.

For the sale of these lambs Martin and I had based the price on the current value of finished stock being sold. I couldn't quibble about it being the going rate – no one was getting any more than this for lambs – but it was still heartbreaking to see them going for such a small amount. It was the same price as I'd have expected to get when I started shepherding eighteen years earlier!

Before the lambs could leave the farm they needed to be ear-tagged, to comply with Ministry regulations, so I set to work early, with Ernie forcing the flock steadily through the handling pens in the shed. It is a relatively quick process if you're dealing with an individual. It's simply a matter of holding the head still and fixing the plastic tag with a specially designed pair of pliers. It only becomes a lengthy job when you are dealing with sheep by the hundreds.

There were other reasons for the slowness with which the sheep were leaving the pens. Like most stockmen, I try to take a pride in the stock that I sell, and the fact that these sheep were destined to form the start of a friend's new breeding flock only added to my determination to make a good job of it. So, as well as ear-tagging the lambs, I took the opportunity to cut the dirty wool from their back ends, and trim up any feet that needed attention.

I still had nearly fifty lambs left to do when I heard the clatter of a heavy vehicle on the drive. Moments later a large green livestock lorry was reversing up to the regular loading spot by the gate that led out of the pens. With unerring accuracy it stopped in just the right place. A few seconds later the door slammed and a familiar figure appeared.

Wilf, one of the area's longest-serving lorry drivers, walked slightly stiffly around the side of lorry, securing his over-trousers as he went, double-checking the position in which it was parked.

'It'll be fine there, Wilf,' I said, still bent double over a lamb. 'I've got a few more to tag, a bit late this morning . . . sorry!'

I felt a little guilty that he, as ever, was bang on time, and I was once again not quite organised.

'You been busy?' I asked him as he followed me back towards the pen of lambs.

'Busy enough,' he replied, obviously understating the truth. 'I was at work at half past four this morning, washing out the lorry. Seem to spend half my life with a power-washer now – that or driving to the disinfection centre.'

He had my every sympathy. Livestock vehicles have always had to be kept clean, but recent events had taken the required levels of hygiene to new heights. Now, before each movement of stock, the lorry or trailer had to be returned to an approved site for washing, before being inspected and sealed with an official metal tag. As Wilf confirmed, drivers seemed to be spending more time washing out than driving.

The sheer burden of this work combined with the deflated state of the industry as a whole had taken its inevitable toll.

'There's not so much work about, but that's just as well really, there's not many drivers left,' he said rather solemnly.

Not so long ago the firm that employed Wilf had run over twenty lorries, serving most of North and mid Devon, but along with just about everyone else in the countryside the previous months had been ruinous for them.

'Boss says the phone never used to stop ringing, but once the foot and mouth hit, it never rang at all,' said Wilf.

He was now one of the few remaining drivers – and thank goodness for that, I thought as I finished off the last couple of tags. After forty years in the job, his knowledge of the location of just about every farm in the north of the county was invaluable – but more than that, his calm gentle approach with livestock made loading and unloading unruly sheep so much easier. I'd experienced more than one lorry driver who loaded stock with what seemed like the maximum of hullabaloo: shouting and whistling and generally stirring the sheep up and making them uncontrollable and wild. Invariably this only succeeded in making the loading process twice as long.

Twenty minutes later I was at last ready. Ernie, displaying a pleasing calmness and control, listened to his instructions and took no time at all to drive each pen of lambs up into the lorry, and then Wilf and I lifted the heavy ramp upwards, and secured it.

Until only a few years ago, that would have been the end of
the job. Nowadays, however, there was still a pile of paperwork
to be completed before the consignment of lambs could go.
Wilf was soon scribbling away, glancing at his watch as he
filled in a long list of questions. The date and place at which
the lorry had last been washed out; the time we started and
finished loading; destination details, holding numbers, flock
numbers – the list went on and on.

I had a copy of the blood-test results to enclose with the
paperwork, and with this safely in his possession Wilf was
finally ready to leave.

'That's it,' he said, handing me his clipboard for my
signature. 'Just sign away your life here, David, and I'll be off.'

Soon the lorry clattered away back up the drive. It was, in
theory, a good moment – two hundred lambs sold; a healthy
cheque heading in my direction in the next few weeks – but in
reality it didn't feel much of a triumph. For the farm to return
to profitability, I needed to see the remaining six hundred
lambs make substantially more money than those I had just
sold. The fact that every sale now required such exhaustive
form filling did little to lighten my mood. I understood fully
why the movement of every livestock animal needed to be
monitored, but the ever-increasing burden of paperwork was a
drain on most farmers, none of whom had much spare time
on their hands.

Heaven only knows what we'd do if there is any truth to the
persistent rumours that European politicians are going to
insist on every individual sheep having its own thirteen-digit
number, which would need to be recorded every time that
animal is moved. According to the politicians, there were
apparently few problems in the trials that they conducted into
using ear tags fitted with these numbers. Clearly none of these
people had ever tried to read five hundred ear tags in the half

dark on a rain-soaked windswept hillside – and none of them seemingly knew that, on average, sheep lose 20 per cent of their tags each year.

This morning's job had taken two and a half hours. If I'd had to read tags as well, it would have taken over four. It didn't bear thinking about. That really would be the end of farming as I knew it, I thought as I trudged back to the kitchen.

CHAPTER FIVE
The End of the Road

It was little wonder the dogs greeted my appearance in the yard with a bout of furious barks and yelps. Wrapped up in a heavy coat and scarf and with most of my head obscured by a black, woollen balaclava, I would have been hard to recognise in bright daylight, let alone in the thin gloom of early morning. It was only when I interrupted their noise with a few familiar shouts of 'That'll do' that they realised I wasn't an intruder.

The winter was placing its unmistakable grip on Borough Farm. The keen winds made the eyes stream and numbed the cheeks and I'd been insulating myself from the biting weather this way for the past few weeks now.

It wasn't a precaution that had been necessary for the dogs.

The dog's resilience, and its resistance to cold in particular, never ceased to amaze me. I remember my first sheepdog, Kim, cooling off once by standing on the ice of a frozen tank. When the ice broke and she fell through, she proceeded to wallow in the freezing water as if there was a summer heat wave. Fern had done something similar recently. Despite her short thin coat, she had swum around in a large water tank on one of the most bitterly cold days of the year. I watched, swathed in my winter woollies, quietly shaking my head in disbelief.

Of course, nothing helped them combat the cold as much as good exercise and today, while I busied myself with the morning feeding, there would be plenty of opportunity for them to keep their circulation going and burn off some of their pent-up energy.

I opened the doors and let the five dogs chase across the yard, scattering a flurry of chaffinches as they sped around. These, together with the farm's population of tree sparrows and a pair of dunnocks, spend much of the winter days picking over any barley that I have spilled. At least they all looked to be doing well on it.

I always started the morning by feeding the lambs that were now occupying a pen at the back of the lambing shed for the winter. There was not much hope of them fattening on grass outside, and I had housed over half of the farm's remaining lambs here. The lack of returns from their sale had affected the way I fed the flock. With little chance of the value of my stock rising, keeping the costs down was beginning to seem like the only way of staying solvent, so their diet now consisted of silage, a nibble of straw and about as much 'crimped barley' as they could eat.

Crimping barley was a new concept for me. Whereas 'normal' grain is not harvested until it has become dry and hard, crimping barley is harvested unripe and soft. The grain is then crushed and rolled in a great heap, before being covered with an airtight plastic sheet. The resulting feed is not only highly palatable and nutritious for the lambs, but also far cheaper. In the past I'd usually purchased my feed from a local mill. This summer, however, I'd bought the crimping barley from a neighbour, and as a result reduced the cost of fattening the lambs by a third.

As I opened the tractor door and climbed up into the cab, a cluster of overexcited dogs formed at the foot of the steps. Obviously the job of tractor assistant was highly prized today. As Fern raised her front feet on to the first tread she was overtaken by the clambering form of Swift, who was soon settling down next to my seat. As she struggled to join us, Fern was soon being challenged by the other three dogs as they jumped up as well.

Flattering as it was to be so popular, there was little enough room for one dog, let alone a quintet. It was as much as I could do to shut the door to keep Greg, Fern, Gail and Ernie out, and as I headed off with Swift, they leapt around in protest.

With the large bucket on the front of the tractor, I spent a few minutes filling up the feed hoppers and laying down some fresh bedding. Over the course of the winter, my sheep will munch their way through one hundred and fifty tons of silage and be bedded on thirty tons of straw. Not many years ago this would all have been shifted by hand, but today farmers can move large bales of hay, straw and silage around easily by tractor. With a pull of a lever, I dropped a bale of straw in the pen with the lambs, and spread it around. There's something rather enjoyable about giving stock clean bedding when the weather outside is uninviting and cold.

With the flock at Borough content and the other four dogs now returned to the kennel, Swift and I headed off on the tractor for the ten-minute drive to the fields at Town Farm, overlooking Mortehoe. The winter's grass was now completely dormant, and the Mortehoe flock were also dependent on their daily food delivery. The tractor bounced heavily over the deeply rutted ground, as I made my way to the heap of silage bales at one end of the field.

Sheep may not be the brightest of creatures in general, but when it comes to recognising feeding time they are geniuses. At the sound of the tractor, fifty or so ewes immediately moved from under the lee of the wall, and headed purposefully forward in anticipation of their breakfast. The constant rain of the previous weeks had taken its toll on this corner of the field: the area of ground around their feeder was now heavily poached with thick mud, and the outdoor flock looked more bedraggled and weather worn than their counterparts under cover back at Borough. The ewes on their spindly legs picked their way through the glue-like mire, their bellies clarted with mud. It did nothing to improve their image.

My old tractor was only two-wheel drive, so lifting a bale with the spike on the front end was prone to make the driving wheels spin. As I lifted a bale of silage from the heap, the tractor began to slip and slide, sending sprays of mud over the back windscreen, and when I moved away from the silage heap in reverse, the wheels were turning at twice the speed of the tractor.

In bad weather, this was nothing unusual and I thought little of it, but when I stopped and slipped into forward gear to turn towards the feeder, I found the traction had gone altogether. The tractor made no movement.

I'd had this tractor for almost six years now and had had problems here and there, but in general it had been a good, reliable and resilient piece of machinery and I was fairly sure it would respond to a little gentle persuasion. In an attempt to get it moving, I tried reverse again, but there was no luck there either. Now I was becoming seriously concerned. If there is one thing that is worse than getting stuck in a tractor, it is getting stuck in a tractor miles from home in the middle of a muddy field in winter.

I slammed the gearbox into forward once more, then quickly into reverse to create a rocking movement. This seemed to produce a few signs of success, so I began shifting between forward and reverse more vigorously. It was here, it turned out, that I'd made my biggest mistake. Almost immediately the gearbox began making a worrying 'graunching' noise with each change. Then, suddenly, there was a violent clunk which reverberated through the whole machine.

Gently this time, I let the clutch out again, but now there was no hint of movement at all.

I shouted in sheer frustration, then climbed down the steps to get a better look at the relevant part of the tractor. The clunk had sounded as if it had come from somewhere within the gearbox. I peered into it, hoping against hope to see something obvious. Of course, there was nothing. I would have to abandon the tractor here and try to get a tow home. I couldn't even finish feeding the sheep. The bale of silage was still sitting high on the spike. While I'd been concentrating on the tractor, the sheep had clustered around. The sweet-smelling scent of the grass had obviously permeated its way through the wrapper and been carried on the wind in their direction. I didn't know what was more frustrating – the fact that the tractor had broken down or that I had no way of getting the silage to them now.

My mood was darkening by the second as I took some string from behind the seat and tied it loosely around Swift's neck. 'Come on,' I said, heading for the gate and the long walk home along the main road.

As we went, I weighed up the options. Tractor repairs are always expensive, and this one was so old it probably wasn't viable to repair it anyway. Even the most basic second-hand replacement would cost several thousand pounds, an expense that we just couldn't afford at the moment, unless we borrowed the money from the bank – but with a third of the farm's income already being spent on rent and bank repayments, it was something I really didn't want to have to do.

I began to wonder whether – a bit like the tractor – we were reaching the end of the road.

CHAPTER SIX

The Changing
of the Guard

The mechanic who called in to look at the tractor confirmed the bad news. It was going to cost more to repair than it was worth. On top of that, it was so old the parts it needed might no longer even exist. So for the time being the only working tractor on the farm belonged to Nick. Sadly his favourite toy wasn't much use during the morning's feeding, although it wasn't ever for the lack of trying on his part.

65

As I bagged up the rations of barley, Nick repeatedly drove his tractor's front bucket into the great heap of grain on the shed floor, scooping up a few cupfuls at a time, tipping them first into a plastic bucket and then a bag. This final part of the process usually resulted in half the barley being spilled on the floor. However, what he lacked in finesse, he more than made up for in determination and by the time I had carried a dozen bags across the shed for the lambs, he had succeeded in filling a bag, which he proudly presented to me.

'That one's for the rams, Dad,' he said, as he attempted to lift the bag into the back of the quad bike trailer.

'Well done, Nick,' I said, helping him haul it up. 'They'll appreciate that after all the hard work they've been doing.'

With the autumn gone, the 'tupping' was over. The rams' work was complete for another year and it was time that they were brought in for the winter.

Rams can be aggressive and awkward customers, even at the best of times. Shedding them, and holding them away from the ewes requires a dog with a lot of determination. For years there had never been a question as to which dogs I would use. Greg and Swift would have been my first choice, without a shadow of a doubt. They formed a perfect partnership that blended Greg's instincts and brains with Swift's courage and ability to obey precise commands, but during the past year time had begun to take its toll on both of them, Greg – now nearing his eighth birthday – in particular.

Greg is a big-framed dog, and I was becoming more and more concerned that the years of relentless running were catching up with him. There was no doubt that he was no longer as quick as he had been in his prime. He was also losing some of the decisiveness that had left the sheep in no doubt as to who was boss. Once it would have been unthinkable to undertake such a job without him, but now I had to balance giving him the work he craved with protecting him from burn out.

As I approached the kennels Greg let out a bark of frustration, almost as if he sensed he was going to be passed over. His tail wagged furiously as I looked towards him.

'Not this morning, Greg. Have a rest and we'll go out this afternoon.'

Instead I chose his old partner Swift, and then Gail, who had now established herself as a reliable dog alongside her mother.

With the two of them, Nick and I headed towards the field directly below the farm, where four Suffolk rams had spent the past six weeks running with a hundred and fifty ewes. It wasn't hard to see that their tour of duty was over. Through the breeding season, each ram wears a harness fitted with a coloured marking crayon across it chest, so that, when a ewe is served, a brightly coloured mark is left behind. A month and a half after I'd first let the rams loose, there weren't any unmarked ewes in the fields.

As Gail and Swift brought the ewes streaming down across the field, three of the four rams hung towards the back of the flock. Over the mating season the rams quite literally wear themselves out. It's not unusual for them to lose a third of their body-weight or, in extreme cases, even work themselves to death. These Suffolks, it seemed, had neither the strength nor the desire to keep up with their ladies. It gave us a head start.

The ewes arrived at pace. I blew a stop whistle to Swift and she came to a halt almost alongside me. At the same time I delivered a long blast to Gail, which brought her around the far side of the sheep, passing the three rams as she went. Before the rams could catch the rest of the flock, I stopped Gail, then a moment later called her to 'Come through', the command for 'shedding'.

Gail's shedding had become highly proficient in recent years, although, like the rest of her work, it was always functional rather than spectacular. She darted in to confront the rams, intercepting them before they could get any closer to their retreating flock. For a second the trio thought about attempting to break through, but it was obvious that they had neither the strength nor desire to try to pass the dog. Gail eyed them, with a look of intense determination, then a second later she was joined by her mother, Swift. The brief resistance was over. The three rams turned away, and trotted obligingly towards the gate.

The fourth ram was a little less willing to be parted from the ewes for another year. By the time we'd safely shed the other three rams, he was already several hundred yards away, trying his best to hide in the midst of his female admirers.

Nick had been watching intently to this point, and now took his own dog whistle from round his neck, put it to his lips and gave it a shrill blast. As yet he had only mastered one, slightly off-tone note. For a split second Gail thought about responding. She took a few quick strides to her left, before pulling up sharply and sending an inquisitive look in my direction. It was enough for Nick.

'Dad, did you see Gail?' he asked, the delight obvious on his face. I could only imagine how I would have felt at Nick's age if I'd got a sheepdog to react like that.

'Go on, blow it again,' I said. He managed a similar sort of sound from the whistle, but this time Gail wasn't to be fooled, and hardly twitched an ear at the tuneless screech.

'Nearly, Nick, but I'd better get them back before they go too far,' I said, ruffling his hair.

This time I sent only Gail to turn the flock back, and left

Swift at my side. If the ewes overshot on their return, they might easily mix again with the three rams, now by the gate.

The sheep trotted obligingly back across the field, until they milled around Nick and me. Slowly between the two dogs, and two shepherds, we managed to manoeuvre the last ram on to the edge of the flock nearest to the gate. Once again, Gail cut in to shed off our target. To keep a single ram away from the entire flock would have been well nigh impossible, so she steered a group of three ewes away at the same time. As the rest of the flock ambled off across the field it was then a comparatively straightforward process to separate the trio of ewes from the ram. With this successfully done, Nick opened the gate and the four rams left the field, and headed towards the paddock.

Nick and I spent the rest of the morning retrieving the other small groups of rams from each field. It turned out to be a rather easier job in the case of one group of six Suffolk rams in the field at the top of the farm. We found them tucked against the hedge line, already separated from the ewes in the flock and standing next to the gate. As a breed these rams have what is known as 'tight wool'. During the cold, wet days of winter, the tight crimps of this type of fleece act like a sponge, chilling the ram and forcing it to burn all its energy just to keep warm. It was clear this group of six had had enough of shivering in the fields for this year. There was no need for any shedding, just a little gentle persuasion to send the weary rabble down the drive.

After lunch, with the rams out of the way, I took the opportunity to move the ewes to the fields at the extreme south-eastern corner of the farm where there was still a little greenness in the pasture. It would be the last chance they'd have to pick over decent grass before the winter really took hold and I'd have to resort to daily feeding with silage. It shouldn't have been a long job, only a short walk for the sheep along the route of the old railway.

The long straight track that runs across the drive of Borough Farm, bisecting the higher fields, was once the route of the railway line from Barnstaple to Ilfracombe, built by the Victorians for holidaymakers visiting the then thriving resort. It was also used to transport the sheep, cattle and rabbits sold at regular livestock markets held at the old Mortehoe Station.

At the point where the tracks crossed our drive, there used to be a level crossing. One of the original 'Stop Look Listen' signs still lay in the back of an old outbuilding at the farm. It had probably been wise advice, given the coastal mists that regularly engulf the area. Like so many rural railways, however, the line had been closed post-Beeching at the start of the 1970s. Since then the track had become a footpath and cycle way – and a very useful throughway along which I could move my sheep.

On Nick's insistence we brought all five dogs with us for the afternoon. I didn't think we needed that many, but didn't want to deflate his enthusiasm. So Greg, Swift, Gail, Fern and Ernie accompanied us to one of the upper fields to collect the ewes.

Sheep tend to be creatures of habit. During the course of their lives they get to know a few routes around the farm and usually want to use only these. The route along the railway track was one we'd taken them along before, but not that often. As we crossed the drive, with the intention of continuing further along the track, they spotted a gate on the right that marked the entrance to another field where several hundred lambs were already grazing. For some reason the leading ewes decided they were supposed to head in this direction and started pressing against the firmly closed gate. A bout of collective stupidity then broke out as, with Herculean determination, the entire flock decided to ignore the open gateway in front of them and press against the closed one to their right. Within a few moments the wood was bending ominously in the middle. Then there was a loud crack as the catch gave up the fight and the gate sprang open.

'Dad, they've gone the wrong way,' Nick shouted, as we both watched the entire flock surging through and bolting across the short stretch of grass between themselves and the lambs.

Panic.

If they got mixed up it would take me hours to sort them out again, back in the sheep-yards. All of a sudden, having all the dogs to hand seemed like a marvellous pre-ordained plan. In an instant five flashes of black and white darted past, and cut across the cussed animals, blocking their path to the lambs.

The effect was just as immediate. With all five dogs fanned out in front of them, the rebel flock took one collective look at the opposition and reluctantly started to beat a retreat. A minute later I had joined in and driven the sheep back through the gate, fastening the badly bent gate-catch as best I could.

Any relief at having shut the gate behind them was short-lived. Once through the gate half a dozen ewes doubled back on themselves, bolting down the drive towards the farm. With the rest of the flock still crammed against the gate and the hedge, it was impossible for the dogs to squeeze by in order to head them off.

Four of the five dogs tried their hardest to wriggle through an improbable gap, but it was Greg, the oldest and slowest of them all, who saved the day. Turning from the sheep, he ran back through the field and alongside the lane. Finding a gap under the fence, he slid beneath. In front of him now stood the tall stone bank, but he soon found a low spot and leapt over it, and into the lane. His arrival was perfectly timed. The escaping ewes had run nearly a hundred yards, but now suddenly found their way barred. They pulled up sharply, before slowly turning back in the direction from which they had come while Greg pushed purposefully on behind them – but even now they weren't quite ready to concede defeat.

The flock turned once more, this time to face the open gate to the railway track, but still they wouldn't go through. A line of ewes faced the opening but leaned backwards, as if they were looking over some invisible precipice. Turning as one to the field gate, they forced it open again and made a second determined effort to join the lambs. Once more the dogs were more than equal to the break, and for a second time forced them back out on to the track.

Clearly the flock needed some leadership here, so this time I caught hold of one of the stubborn creatures, and pulled it through the gap and along the track while uttering my best bleating noises. It may have sounded surreal, but it did the trick. It's no accident that people use the expression 'following like sheep'. Within moments the rest of the flock were trotting along the correct track as if they'd known it was there all along.

'They are silly sometimes,' Nick said in a matter of fact sort of way. 'It's a good job I told you to bring all of the dogs, isn't it?'

With the setbacks of recent weeks, the dogs had been far from the forefront of my mind. As I finished off in the yard that morning, however, I found myself taking stock.

There's an old saying in the sheep-farming world: 'There's no good flock without a good shepherd, and there's no good shepherd without good dogs.' From my earliest days as a shepherd, it was a piece of wisdom that had always rung true.

My first dog, Kim, had shown me how much of an asset a good dog could be. It wasn't just the benefits she brought me either. The lives of the sheep became far less stressful when they were handled with the subtle force a good sheepdog uses. In the twenty or so years since then, my belief in the wisdom of these words had only deepened. There were times – like this morning – when I felt a mixture of awe and gratitude towards my dogs. I knew I'd be no good as a shepherd without them.

To see skills like those Gail and Swift had displayed this morning, and the entire pack had shown with the ewes this afternoon, was especially pleasing for me. It takes a long time to train a dog to work to that standard. It was something I'd been able to test at the sheepdog trials in which I'd run both the older dogs, but I find it far more rewarding to see those skills being put into practice in the 'real world' of day-to-day shepherding.

And yet, as I looked at my quintet of dogs, there was a sense that the pecking order within the pack was about to shift. Greg was clearly approaching the age at which he would increasingly take a back seat on the farm. He would undoubtedly cope with another season, but I knew I'd soon have to phase him out of working life. Swift still had a few more years of work in her, but she too couldn't go on for ever. The progress the younger dogs had made in the past couple of years was hugely encouraging, but, as I thought about the younger members and compared them to the 'old guard', I could see there would have to be some changes. I would miss Greg in particular. For all their talents, none of the younger members had his wisdom and in all probability I would never find the like of him again. As he'd proven once more this afternoon, he had a vision and an intelligence that was rare indeed. Swift too would be a hard act to follow.

This evening, as I let them have a final run before shutting them in their kennels, I realised I would have to start thinking about finding a new generation of dogs, and a successor to Greg in particular. I also realised I would have to do it soon. When it comes to keeping a team of sheepdogs at the peak of their game, you need to think a couple of years ahead. By the time another pup was trained up, Greg would be in retirement, and even Swift would be heading that way.

Getting a new dog was always a family decision as much as a farming one. I'd have to talk to Debbie about the idea, as she inevitably ended up having to look after a new dog, especially while it was young. She was the most likely to object to the extra work a new dog brings. Trying to sell the idea of a new pup to the children, on the other hand, was about as difficult as trying to convince them to eat chocolate. In fact if I mentioned it to them first, it was almost guaranteed that they would begin pestering Debbie and me so relentlessly we'd have no choice but to get one.

Not a bad plan, I thought.

***

The vet gently held the lump on Swift's foot between his forefinger and thumb, then peered over his half-moon spectacles to deliver the bad news.

'It'll have to come off, I'm afraid,' he said, looking once more at the small, pea-sized swelling on the side of the foot. 'And we'll probably have to take the toe along with it.'

I'd been nursing a bad feeling about this lump since I'd spotted it the previous day. At first I'd put the strawberry coloured bulge down to infection from a thorn, a common cause of grief for dogs at this time of the year when hedges are trimmed. On closer inspection, however, the lump appeared more solid than that and didn't look as if it could be easily lanced either. Debbie and I had prodded and poked at it for several minutes without coming up with a diagnosis. Swift didn't seem to be in any pain, but it didn't look like a problem to be ignored.

'Looks like another trip to the vet,' Debbie had concluded with a sigh.

While I'd been half expecting bad news, it was still mildly shocking to hear the vet confirm our worst fears. My main concern now, obviously, was whether it was life threatening.

'Do you think it's malignant?' I asked, rubbing Swift's nose at the same time.

The vet was still concentrating on the lump and thought for a while before answering, 'There's a good chance it's cancerous, but we shan't know if it's malignant until we have it analysed.' He could clearly see the concern spreading across my face. 'But the good news is that cancers on the feet and legs don't usually spread.'

'There's a chance though?' I asked hesitantly.

'There's always a chance with a lump like that, but we'll know more once the results come back,' he said. 'The best thing to do is to get on and operate. I can do it this afternoon if you leave her with me.'

I've never considered myself overly sentimental about my dogs, but as I led Swift along the rows of wire-fronted kennels at the back of the surgery, I couldn't help but wonder what she must be thinking. On either side a collection of pet dogs whined, barked and wailed, all of which added to her apprehensive demeanour. Then, when the vet-nurse finally found an empty cage and shut the door behind her, the sorrowful look that she gave me through the grille, tugged at the heartstrings.

'Good girl, stop there, Swift,' I said to her gently. Her tail hardly twitched. She almost certainly understood that being left at the vet's was not good news, and perhaps, just as importantly to her, she also realised that I would be going to work without her.

'You'll be all right, Swift,' I called as I turned away and left her – but as I drove back to the farm, I couldn't help wondering whether her best days were now behind her.

I spent the rest of the morning and early afternoon working on the hedging and fencing – I had made little progress in recent weeks, and at this rate there seemed little chance of finishing the work before lambing – but as I worked my mind kept wandering back to Swift.

Midway through the afternoon, I could concentrate no longer, and headed back to the vet's where I collected a very sleepy, still half-anaesthetised, but successfully treated Swift.

'All went fine,' the vet-nurse reassured me as I scooped Swift up, being careful not to touch the heavily bandaged foot. 'Give us a ring next week and we should have the results of the test.'

Back home, I carried Swift carefully into the kitchen and placed her in a box by the window. There was nothing more to do than leave her to sleep, so I went off in search of the rest of the family.

With darkness already beginning to set in, I found Debbie, Nick and Laura helping Clare to retrieve the ducks from the small pond in the field near the house. The girls had hatched fifteen ducklings in an incubator during the spring. As treasured members of the menagerie, they couldn't be left in the open overnight for fear of the fox. Usually the ducks would trot back to their house, with nothing more than the offer of a little corn. Occasionally, however, they would stay out until darkness fell and become reluctant to leave the sanctuary of the pond.

'How's Swift? Have you collected her yet?' Debbie shouted, spotting me in the gloom.

'She's in the kitchen. She looks OK, but we shan't know what the results are from the lump for a few days yet.'

Clare and Laura pricked up their ears at this news of Swift's return and headed off towards the house to see to the wounded warrior. Only Nick hung behind, asking questions nineteen to the dozen.

'Is Swift going to be all right, Dad? What did the vet do to her? Will she have to go back to the hospital? Does her leg hurt a lot?'

It was obvious he was finding it all a little hard to take in, but it was clear too that, to his young mind, it was distressing to think of a member of his extended family having to undergo such an operation.

By the time Debbie and I reached the house with Nick, we had managed to convince him that Swift would probably make a full recovery and he looked a little chirpier. The sight of Swift lying semi-conscious with a large bandage on her leg soon sent him back into his worried shell, however. While Clare and Laura stroked Swift lovingly, he hung back, looking rather warily at the familiar figure now lying in such an unfamiliar state.

My heart went out to him.

For the children's sake as well as my own, I hoped the vet's prognosis had not been unduly optimistic.

———

By Christmas morning the mantelpiece in the living room was overflowing with cards. One in particular had been given pride of place. It was a suitably festive picture of a shepherd at work in an idyllic, snow-filled winter's scene, feeding hay to his flock.

'It's you, Daddy,' Nick had told me excitedly when it had first been pointed out to him.

If only.

As the big day arrived I couldn't help wishing Borough

Farm had been given a dusting of snow to match the image on the mantelpiece. The prospect of crunching my way over frozen-solid ground would have been far more enticing than another morning wallowing around in the mud. I went out as early as possible, so that I could be back in time for church and the rest of the Christmas Day routine. I left the children in the living room, where they were already playing, wearing or eating the assorted presents Debbie had spent the past month accumulating. Nick, needless to say, was most excited about the latest addition to his miniature tractor collection, a JCB digger.

Outside, it was mild but murky. The hazy mist hanging over the valley looked set to stay there all day. With Swift still recuperating in the house, I took all four fit dogs with me as I did the minimal amount of work necessary, feeding the ewes under cover and zipping around the lower fields on the quad bike to check on the flock there. I had given the Mortehoe flock plenty of silage late on Christmas Eve so, just for today, I could leave them alone. Greg, Gail, Fern and Ernie returned to their runs to find that the children had wrapped them a chew each. They showed their appreciation by immediately devouring the lot, wrapping paper and all.

At ten past ten, with the children – and me – washed and dressed up, we were all ready to make the lightning dash to church, three or so miles away. The picturesque, Georgeham church is always busy on Christmas morning and today was no exception. Once more it was standing room only by the time we arrived.

The local farming community is usually well represented at the Christmas morning service, but today it seemed that all our neighbours had finished their morning milking or stock feeding early so they could attend. The sense of the community having come together was real – and rather warming, to me at least. The evocative sounds of the organ, the local school choir and the familiar strains of Christmas hymns were just as uplifting.

When the service was over I shook hands and exchanged Christmas greetings with some of the throng as we filed from the church. For the Cook family, who farmed nearby, Christmas Day marks a welcome turning point of the winter. Their big business at this time of the year is growing Brussels sprouts. They looked delighted that the annual eight weeks of picking, grading and bagging tons of them was at an end once more.

As we chatted there was hardly a mention of the events of earlier in the year, just a few quietly spoken words. 'Let's hope the New Year brings some better luck,' someone said. There had probably been many a prayer murmured along those lines this morning.

CHAPTER SEVEN

Facing the Future

The first week of the New Year brought a rare sight at
Borough Farm – frost. In the sheltered lee of the trees,
where the overnight air hung still, a crisp white blanket
covered the grass. The lower branches of the overhanging ash
were coated with a thin layer of ice that glistened in the
morning sun.

It being a Saturday, the children busied themselves around
the yard. Nick was at the steering wheel of his tractor, clearing
a sizeable pile of gravel and mud from the yard drain. Laura,
clad in coat, gloves and pink woolly hat, had intended to clear

out her guinea-pig run, but had had her plans hijacked by Clare. In general she didn't share her big sister's 'horsey' interests, but every now and then Clare would succeed in bribing her into helping her with dunging out the stables, or cleaning the tack. Today the 'bribe' was a ride on Minty, which was rather bizarre as Laura had very limited interest in riding.

I now watched with some concern, as a rather terrified-looking Laura was led up the drive on Minty's back, bolt upright, staring straight ahead, somewhat reminiscent of the last standing skittle on a bowling alley. Whatever Clare was getting in exchange, it was clear she had had the better of the bargain.

The frost was timely. It isn't often that the ground dries enough over the winter months to allow me to work in the woods without becoming embroiled in a sea of mud. So, with the sheep checked, I headed down the slope to Borough Valley and the newly fallen tree that had presented itself as a candidate to be sawn up for winter logs.

The roar of the chainsaw destroyed the peace of the woodland in an instant. Two wood pigeons clattered their way from the top of a nearby sycamore, but it was the sight at the top of the bank that caught my eye. A red deer stag bolted from amid a patch of thick bramble. I watched him as he then sprang majestically over the bank at the higher reaches of the woods and disappeared from sight.

We have a small transient deer population in the Borough woods. Judging by the size of his antlers this was a young stag and he would be unlikely to be on his own, so I watched in case there was a second, but there was no sign.

A few hours with a chainsaw can cut an astonishing amount of wood. As the light began to fade, a healthy looking pile of logs had accumulated around me. There was easily enough for what was left of the winter, so, quietly pleased with the day's work, I began to make my way back.

The walk back up to the farm offers some of the most compelling views of this corner of North Devon. Too often I took it for granted, but today for some reason I found myself pausing at the top of the hill, looking northwards towards the sea. With the sun fading, the valley's population of rooks were settling for the night. Far below, a great black plume of noisy cawing birds squabbled their way high into the evening sky, then swirled for a few moments before descending into the silhouettes of the distant beech trees.

Rookery Wood has been marked on the map of the area for centuries. For generations of farmers at Borough Farm, the sound of the birds would have been familiar. With so much uncertainty now, here at least was something that was completely immune to change.

Back in the yard, I'd just scooped up a bucket of water from the butt when a chorus of barking from the direction of the kennels alerted me to a new visitor. Almost immediately I recognised the ruddy-faced figure emerging from his pick-up truck. It was another local farmer, Colin, whom I'd not seen in an age.

'How are you, Davey me boy?' he bellowed, a broad grin on his face, as he held out a hand so worn and hardened it felt like a wicket-keeper's ancient glove. It wasn't just his hands. Everything about Colin told the story of a life of hard grind. His boots were so worn the steel toes were poking through the leather, and even his cloth cap was ripped along one side, revealing a clump of greying hair. Yet he had an air of unquenchable optimism about him.

Farming can be a solitary occupation, and for the most part that more than suits me. It's not that I avoid passing the time of day with farming neighbours – I always enjoy a chat with a colleague – but there rarely seems to be the time or opportunity, and during the worst of the foot and mouth outbreak there was also the risk of contamination to consider. Such was the siege mentality, I remember that when one friend pulled up at the top of the drive to discuss the latest rumours of the disease's spread, I had to stop myself from telling him to keep away in case he was unwittingly carrying the infection.

Thankfully that paranoid period had drawn to a close and I enjoyed a visit from someone who shared the same life. I still found it rather touching that 'old timers' like Colin seemed to accept me as one of them. Debbie and I had only moved to Mortehoe a decade earlier, after all, and therefore came nowhere near to fulfilling the commonly held criterion for being a true local – 'three generations in the parish graveyard'. That had never mattered to Colin. He respected what I was doing at Borough Farm, and I felt the same way about him.

As we struck up a conversation about farming staples like the weather and the number of lambs about, I was careful to avoid a direct mention of last April's events. Colin's farm, only five miles away, had been hit as hard as any in the county. He'd lost a dairy herd and a herd of fattening cattle, as well as a flock of nine hundred sheep plus their lambs. However, it was inevitable that the conversation would veer towards what had, after all, been the dominant event not just of the past year, but probably the past decade. When it did, the grin subsided from Colin's face and the life seemed to drain from his eyes.

'Never again,' he said looking down at the ground for a second.

He didn't remain downcast for long, however. Within a few moments, the grin was back as he remembered the reason for his visit: 'Wanted to talk to you about a new dog. Our old dog, he's got to nearly nine now and I'll never cope if we have a thousand ewes again next spring. 'Ave you got any puppies you'd be willin' to part with?'

Slowly my sheepdogs were beginning to gain a reputation locally. Greg in particular was being used with some frequency as a stud dog.

I'd have liked to be able to help Colin, but I couldn't.

'I haven't at the moment,' I replied. 'To be honest, I've not really thought too much about breeding dogs recently. I thought people might not be in the mood for new puppies about the place.'

Colin looked at me with a slight degree of surprise. 'There'll always be people wanting a good dog,' he said, shaking his head gently. 'So long as there's sheep 'bout.'

Although there were already a good number who had restocked their farms, a lot of others were turning their back on the farming industry.

'Lots of people won't go back into sheep after they've been culled though, will they?' I said, the growing sense of defeatism apparent in my voice.

I was a little taken aback by the strength Colin's conviction.

'Course they will, course they will. I said to my missus, "The day I don't want to get out of bed in the morning, 'tis the day that I give up farming." It hasn't happened yet.' Colin farms with his two sons. 'After they'd taken our stock, I asked my two boys straight: did they want to go back? They both said "Yes" straight away.' The pride was evident in his voice.

He looked at me and seemed to be weighing up my thoughts. ''Tis in the blood, Davey. Yours and mine. What happened is in the past. Now it's time to think about the future.'

We talked for a few more minutes, about sheepdogs. I said perhaps I could breed a litter this spring, and if I did, Colin would certainly get a pup.

'That'd be decent of you, Davey,' he said. 'Give me a ring when you've got a nice one you think'll suit me.'

He was just about to head back across the yard when he spotted a group of lambs in a corner pen. 'Sheep are looking well,' he said.

'They're all right now, but we had some pasteurella back during that filthy weather at the end of the year and they weren't so good then. The vet reckoned I should give them a dose of penicillin, but I thought it might do more harm than good, sticking them in a pen and stressing them out.'

'We've all had it, Davey,' Colin said, readjusting the cap on his head, and leaning into the pen where my lambs were gathered. 'A few years ago I lost any number. You did the right thing not chasing them round in a pen. It stops on its own in the end.'

A few moments later I watched him climb back into his truck, and head off up the drive, his old dog barking gamely through the mesh in the back of the pick-up as he went.

It's strange how an unexpected visitor can make such an impact.

Perhaps it was the walk back up from the woods that had put me in a vaguely philosophical mood, but as I tied up the day's work in the sheds, Colin's words started replaying in my mind.

As I'd dealt with the pasteurella outbreak I'd thought long and hard about whether my course of action – or inaction – had been right. Colin had backed up my approach and that alone made me feel a little better. He'd also made me realise how valuable an asset I had in the dogs. I'd put a lot of time and effort into training and breeding my pack and it was pleasing to think that was recognised by such a knowledgeable peer, and that it might earn me a few valuable pounds when things got back to some semblance of normality. It wasn't just selling puppies to farmers like Colin, either. Before foot and mouth I'd run a few successful training courses for other farmers and their sheepdogs. It had been all too easy to forget about that part of my life in the past months. He'd reminded me of what I had there.

But it was his determination to move on from the dark days that made the most impact on me that afternoon. The last year had been difficult, and there was always a chance that things would get worse before they got better, but in comparison with many local farmers we had been lucky. If Colin could take such a positive attitude then so could I. He was right, it was time to look to the future.

If I'm honest, I avoid towns like the plague. I much prefer the peace and relative calm of the countryside. As I drove through the outskirts of the nearest sizeable town, Barnstaple, my aversion was, if anything, growing even stronger. I really wasn't looking forward to the morning ahead.

I had called the bank a week or so after the tractor had died on me. Whatever our financial situation, we simply couldn't go through the winter without one. As I stood in the lift heading up to the first floor offices, I was finding it hard to be positive. I'd been successful in getting money in the past, but asking for a loan to buy a piece of land or some more stock was a relatively straightforward matter. It's always going to be easier to convince the bank to lend if there is a chance that you will earn something from your prospective acquisition.

My bank manager Bill greeted me cordially, but there was no mistaking the businesslike tone of his voice.

'So, what can I do for you, David?' he said, drawing a pen from his pocket and preparing to take notes. The night before I'd prepared a few cash-flow sheets. They were, of course, on the optimistic side, but I felt they made a semi-convincing case. I talked Bill through the figures and told him I thought they would comfortably allow me the extension of the overdraft I was asking for in order to buy a replacement tractor.

I had come to know Bill well over the previous few years, and had found that he was largely supportive of our farming efforts. I also knew that banks were beginning to feel uneasy about the outlook for farmers: there was only so much support they could give.

His pause was a little too long for my liking. For a moment or two I was sure he was going to turn me down.

'OK, David, you can have the money this time,' he eventually said, his tone making it clear there was going to be a caveat, 'but I must tell you, unless things take a turn for the better, it will be the last time that I can say yes.' He went on to explain that he was getting a lot of pressure from his head office and that new directives on lending were coming into force: 'Things look pretty bleak, if I'm honest.'

I knew that I was not alone in facing an almost daily battle for survival. Bill painted a grim picture of farmers whose income had been cut in half and who had been left with huge debts. 'The bank's policy is to be sympathetic, but in several cases the only solution has been for the customer to sell up,' he said, with an air of sadness.

I wasn't sure if it was meant to make me feel better to know that other farmers were in such a desperate plight. For now, however, I was only concerned with my own predicament, and I was hugely relieved to have been given the go ahead to get another tractor.

Bill picked up on my relief and as he did so his mood lightened a little. We chatted for a while longer, then I headed back down in the lift. Even the weather had improved during our meeting and the sun was trying its best to break through a bank of clouds over the north coast. There had been plenty of times in recent weeks when I'd wondered whether the time was approaching when I would have to think seriously about retraining and a new career. The idea was almost inconceivable, but there were times when it seemed the only way I could ensure Debbie and I didn't lose our house and way of life.

Now, at least, I felt I'd been given a chance to fight another day.

Perhaps we'd turned a corner. Perhaps this was the beginning of that future I'd felt so optimistic about a few days ago.

'Clean tractor, low hours, done little work.' The advert in the *Farm Trader* had looked promising enough for me to mark it in bold felt-tip pen. As I picked up the phone to ring the accompanying telephone number, I had a good feeling that perhaps my search for a replacement tractor was over almost before it had begun. I'd managed for a while, but the lack of a tractor had become a real problem, and for the previous couple of weeks I'd resorted to paying my nearest neighbour, Adrian, to dole out the daily rations of silage using his. We were doing each other a favour. Like many small farmers, Adrian supplemented the income from his dairy cows with contract work, when time allowed. He helped me out most springs, mowing the silage grounds and applying fertiliser.

Yet I knew this arrangement couldn't continue for much longer – for one thing it didn't make economic sense. So, with the extension on my overdraft finally approved and available to be drawn on, I'd started the hunt for a replacement machine. It had been Adrian who'd suggested I got a copy of this week's

Farm Trader. Just as he'd said, there was a good selection of second-hand machines, everything from balers and muckspreaders to combine harvesters and tractors. The one I'd circled in pen certainly seemed to fit the bill. It was the right horse power, it was ten years old, but apparently hadn't done much work. The voice on the other end of the line was encouragingly friendly.

'Oh yes, the Ford,' he replied. 'A tidy machine. She's done less work in ten years than most do in one.'

This sounded even more promising, I thought. I'd had good experiences with Fords in the past, and a ten-year-old machine was likely to fall into my price range as well. I did my best not to sound too enthusiastic, though.

'I'm in North Devon. Whereabouts are you?' I asked, hoping he'd be somewhere in the southwest.

'Wiltshire,' he replied to my consternation. 'Not too far.'

'Not too far?' I said to myself. 'That's half a day's drive in the Land Rover.' It sounded like an opportunity too good to miss, though, so I bit the bullet. 'All right, are you about tomorrow? I'll come over and have a look.'

———

As the Land Rover rolled along an endless dirt track, under the shadow of a railway embankment, I was beginning to have my doubts about the wisdom of my trip. The location I'd been directed to by the man on the phone looked like a derelict junkyard. To the left even the steepling grass and brambles couldn't obscure the remains of rusted farming implements, long since abandoned. What really bothered me, however, was the collection of tractors on the right-hand side of the track. Each of them looked like they'd been standing there for an eternity. Many were missing wheels, some were missing engines and a few were missing both. The only positive point was that the further down the track I went, the less decrepit the machines became – but that wasn't saying much. Perhaps, I thought, this was where tractors came to die.

Finally the track cast around to the left, and I drove through a pair of tall steel gates, pulling up alongside a grey Portakabin that stood amidst a stack of hydraulic rams and a massive heap of blackened engines.

It didn't seem the sort of place to knock politely, so I banged loudly on the cabin door with my fist. Somewhere to the back of the shed beyond it, what sounded like a large, heavy-smoking dog barked hoarsely. A few seconds later a large steel door on the side of the shed slid open, and a broad man, with a massive oily beard and a pony tail, sloped forward from the gloom within. He paused at the door while I walked towards him, and leaned on the frame in a slightly intimidating way. I decided it was best to cut to the chase.

'You had a tractor for sale in the paper this week. I rang up yesterday. The name's Kennard.'

His face hardly flickered; he just continued to stare with his hard black eyes. When he eventually opened his mouth, the voice that emerged bore no relation to the friendly, enthusiastic one I'd spoken to on the phone twenty-four hours earlier.

'What tractor you after?' he asked, his mouth hardly moving behind his great beard, eyeing me up from head to toe as he spoke. He then nodded towards the collection of rusting antiquity that stood by the side of the drive. 'You passed all we got on the way in.'

My heart sank. I'd seen nothing that looked like an improvement on the tractor back in my yard. In fact mine looked less like a candidate for the scrapheap than these.

'I rang up about a Ford, low hours on the clock, ten years old. The bloke on the phone said it was a tidy tractor,' I said, with a dismissive nod at the machines behind me.

My words were once more met with a blank look. I'd quickly come to the view that he had absolutely no interest, and no intention of helping.

'You can have a look if you like; there's a Ford down there; don't 'spect it will start though,' was about all he could manage.

'Doesn't sound like the one advertised,' I protested. 'The bloke on the phone said it was a decent tractor, not a non-runner.' I was beginning to feel a little exasperation. 'I've driven a long way.'

He gave me an inquisitive look then grunted, 'Where from?'

'North Devon.'

'That's not far. Come back when the guv'nor's 'ere,' he muttered, retreating into the grimy depths of the shed.

'But I . . .'

It was no good. The door screeched along its runners and banged shut. I just stood there fuming. Then I stomped back to the Land Rover and slammed its door in futile response.

Calming myself, I remembered that there was indeed one Ford tractor among the clutter that lined the drive. I might just as well stop and look. 'Appearances can be deceptive,' I told myself, clutching hopelessly at straws.

A close inspection soon punctured my fragile optimism. One of the tractor's back tyres was flat and perished to the point of splitting; only a few panes of glass remained – the rest had been either removed completely or else shattered so that jagged edges poked out of the frames; a particularly stout bramble had grown up through where one of the lower front windows should have been, and had also woven itself around the steering column and wheel. If it had any scrap value, it wasn't much.

'Clean tractor, low hours, done little work,' I muttered to myself ironically, as I climbed from the step and pushed my way back through the long grass towards the Land Rover.

I was just about to drive back up to the shed, for another attempt to get some sense from the oily gorilla, when a pick-up truck appeared. As it pulled alongside, the window was wound down to reveal a slightly older figure, with greying hair beneath a trilby hat, and a far more affable expression on his face.

'You wouldn't be the bloke lookin' for a Ford tractor, would you?'

My spirits rose instantly. It was the man I'd spoken to on the phone: 'the guv'nor', as the gorilla had referred to him.

'That's right,' I said, mustering the first smile of the morning.

'Well, you won't find it there,' he said, nodding towards the bramble patch behind me.

'The chap in the shed said that this was the only Ford on the place.'

'John? Well, he wouldn't know, would he?'

'I was just about to head for home, give it up as a wasted journey.'

'Well, you won't have to do that, not if it's a tidy little tractor you're after. Jump in and I'll take you to see it,' he said, swinging open the passenger door.

As we headed back down the track he introduced himself as Ronnie. 'Loaned the tractor out to a customer, while we repair his machine. It's only just a mile or two away,' he went on by way of explanation.

It was the longest two miles I've ever travelled. After twenty minutes squeezing our way through some of Wiltshire's narrowest lanes, I was beginning to feel that this might just be some ploy to get me out of the way, while 'oily John' turned my Land Rover into scrap metal, but finally we rounded the corner into a small stone-built farmyard.

Ronnie had spent the journey extolling the virtues of the tractor.

'Never had a day's hard work in its life,' he said. 'If it wasn't ten years old t'would only just be run in.'

The tractor that had drawn me all this way stood in the corner of an open-fronted barn. I couldn't help noticing it had a power-harrow attached, which wasn't a great sign. The power-harrow is perhaps one of the most power-sapping of all farm machines, and the one on the back of my prospective tractor was far too big to belong on such a low horse-powered machine.

I wasn't exactly surprised to hear Ronnie dismiss this. 'He's just bin moving the harrows across the yard, I shouldn't wonder,' he said with a cursory wave of his hand.

Ronnie soon disappeared in search of the farm's owner, leaving me free to give the tractor a thorough inspection, so I started it up. The engine and transmission were, as far as I could determine, in good order. So too was every other working part. Indeed, the whole appearance of the tractor was just as Ronnie had said – 'clean'. By the time he returned, fully half an hour later, I'd come to the conclusion that this was actually an excellent piece of equipment.

From a negotiating point of view this was, of course, not good. I was going to find it hard to barter a discount.

'So what do you think, Mr Kennard? A tidy tractor, don't you agree? All yours for a little over four thousand pounds,' he said, getting right down to business.

I looked at him, sucked heavily through my teeth, then returned my gaze to the coveted machine, trying to look serious.

'It's a bit old,' I said, unconvincingly. 'Don't know how many years it'll have left in it. It's just the age when things go wrong.'

'Done less work in ten years than most do in one,' Ronnie responded, repeating the line he'd used on the phone yesterday. He then went off into a well-rehearsed sales spiel which was soon drifting over me as I stood, hands in pockets, doing my best to look sceptical. It was hopeless. It was just the tractor I was looking for, and Ronnie knew it.

After another couple of minutes of shadow boxing we eventually settled on a price. I got him to agree this included delivery to North Devon, and managed to knock off a miserly two hundred pounds, but I came out of it feeling that the deal had definitely favoured Ronnie rather than me.

Still, as I started on the long drive back, I told myself I had the right tractor for the job. It was a little old, but on the positive side it lacked the electronic gadgetry that newer tractors are loaded with and which can deliver crippling repair bills as a machine ages. I headed home, if not feeling pleased with the day out, then at least feeling that I had solved one problem and, in my own mind at least, made an important statement. Buying what my accounts would record as a 'capital item' was a sign that Borough Farm was not in the process of folding. Far from it, in fact.

As I swung open the kitchen door, the air was warm and suffused with the smell of a casserole cooking in the Aga.

'How did you get on?' Debbie asked.

'Well, I bought a tractor, so I suppose I must have got on all right,' I said, not really sure as to whether spending all that money represented a good day out. 'They're going to deliver it the day after tomorrow.'

Nick's ears pricked up immediately as I said the magic 't' word. 'Dad, did you say you're getting a new tractor?' he said, barely able to contain his excitement. 'Is it four-wheel drive?'

Setting off for Wiltshire that morning, I'd resisted telling Nick I was going to look at a replacement tractor for fear of disappointing him if I returned empty handed. The news that we were going to get a machine lacking the maximum traction would be a more minor letdown, I reckoned.

'Sorry, Nick, it's only two-wheel drive.'

'Oh,' he said, looking crestfallen for a moment. 'Maybe we'll get a four-wheel drive next time.'

If he was disappointed, he didn't remain that way for long. Within a minute or two he reappeared, having found one of his collection of model tractors from his toy box and was pushing it carefully across the floor, chugging contentedly to himself.

CHAPTER EIGHT

A Nasty Itch

'As the days grow longer, the cold grows stronger.' The old adage certainly carried a ring of truth this year. The late January evenings were slowly beginning to draw out, but the temperatures had barely risen at all. Spring was still a distant prospect.

At Borough Farm, the ewes had come through the worst of the winter well. They seemed to be thriving. If only the picture had been so rosy at Mortehoe.

Several times during my recent visits I'd noticed an odd ewe turning its head and pulling at the wool on its flank with its teeth. The numbers hadn't been significant – only a few of the two-hundred-and-fifty-strong flock had been doing it – but it was becoming more and more of a cause for alarm. Some of these sheep were now showing a cotton-wool-like area on their sides where they had been 'nabbing'. There were several potential explanations for this, but one obvious and deeply worrying one was sheep scab.

I'm not sure why, but sheep scab is the ailment that carries the greatest stigma among sheep farmers. It's probably got something to do with the pride that most of us take in the welfare of our flocks. Scab seems to wound that pride deeply, perhaps because we imagine others will see it as a sign of neglect or irresponsibility, or will think we were too tight-fisted to run our flocks through a sheep dip during the summer.

The truth is, however, that farmers can never protect their flocks 100 per cent from the problem. The parasite is spread by new stock brought on to the farm from market, or can be introduced by wandering sheep. It was the latter possibility that was really worrying me. Although the wanderers I'd picked up off the road at the end of last year hadn't shown any obvious sign of sheep scab, it didn't mean they hadn't been carriers – and if they had been, then I was probably responsible for disseminating the parasite out to Mortehoe. All I'd have had to do was carry one single, microscopic mite out there on my clothing. Once it was established it would flourish and multiply at terrifying speed until it infected the entire flock.

The rate with which sheep scab spreads is what really identifies the disease. Every sheep can be rubbing and biting itself to distraction within a couple of weeks, once a flock is infected. Which was why I had my doubts. In the fortnight since I'd spotted the nabbing, the numbers hadn't swollen at all. If I was honest with myself, I'd been hoping it would conveniently disappear, but, as I headed out to Mortehoe this morning, I knew that with lambing coming up soon there was no way of ducking it any more.

The problem is best treated by dipping all sheep in a bath of organophosphate solution, or giving every sheep a small injection of Avermectin. I wanted to avoid dipping if I possibly could. Apart from the stress caused by throwing heavily pregnant ewes into a bath of cold water at this time of the year, dipping is seriously unhealthy for humans as well. There is an ever-mounting body of evidence to suggest that the chemicals used cause long-term illness to people who are exposed to it over prolonged periods. However, the injection wasn't without its downside either. Avermectin is highly effective as long as every sheep is given the correct dose, then moved away from the fields where the infection occurred, but its cost is considerable. To give every ewe an injection would come to several hundred pounds, possibly more than a thousand. I worked the precise figures out in my head and concluded that it would eat up around 15 per cent of the small profit I made on a sheep in a good year. This year, however, the profit was likely to be so low that the figure might be nearer to 25 per cent of the proceeds of the entire Mortehoe flock. It was a huge chunk to take out of my income.

An old, Kentish stockman once gave me a very wise piece of advice: 'The only way you can really tell what's going on with your stock is to sit and watch them for a while.' This is easier said than done. The daily routine of feeding, moving and treating sheep provides plenty of opportunity to cast a quick eye over the flock, and ordinarily this is more than enough to spot the more obvious problems. When the problem is subtler and less easy to detect, however, these situations aren't ideal. Sheep are normally distracted when they are interacting with the shepherd and his dogs. It is only when they are relaxed and free to do as they please that their behaviour reverts to normality. It is then that you have the best chance of spotting the less obvious problems. Usually there's no time to do this, but, with the itchy ewe situation getting more and more frustrating, I'd decided to make some this morning.

The problem had now become a real puzzle. Each time I'd inspected them in the past month or so the result had always been similar: the same number of ewes displaying the same symptoms. A few were beginning to look a little scraggy, with a sizeable patch of wool hanging from the midriff, but none of them were displaying the telltale signs of true sheep-scab infestation – a frantic rubbing and scratching of themselves.

I found myself a good vantage point and watched the hundred or more sheep Greg, Swift and Gail had brought in. I signalled for the dogs to come back to me, then sat and waited.

Slowly the flock began to turn their attention to grazing once more. With heads down they began to disperse, picking their way from one swathe of grass to another. As they did so the extent of the problem became obvious. A ewe turned 'nabbing' just for a second or two, as she made her way to drink at the spring. She was followed a minute later by a second sheep doing the same thing. Then another rubbed herself against a rock that protruded from the gorse. Over the course of ten minutes I watched perhaps twenty sheep showing the same signs of irritation.

This was the proof I'd needed – but least wanted to see. I still had my doubts that it really was scab. If it was, then why hadn't it spread like wildfire? Whatever the cause, however, it was now clear that this situation had to be tackled. If it wasn't resolved before lambing, then the youngsters would become infected soon after birth, meaning that they too would require treatment. Before I made any final decision, I decided to seek a second opinion.

———

'Looks like scab to me,' the young vet announced matter-of-factly, seemingly oblivious to the implications of his diagnosis, 'but let's put it under the microscope to make sure.'

Finding myself back at the vet for the third time in four months seemed like a minor disaster. Swift's lump was unavoidable, but to need a consultation on a second ovine problem in that time left me feeling despondent and embarrassed, as the vet had probably gleaned from my rather defensive diagnosis of the situation.

'Well, it's certainly not behaving like scab,' I'd explained to the white-coated young man, one of the ever-growing throng of new, junior vets that seemed to be employed at the surgery. 'It's been going on for six weeks now. It hasn't got any better, but it hasn't got much worse either. If it really is scab, then the whole flock should have it by now, shouldn't they? Surely there should be wool dropping out everywhere?'

The vet had looked at me rather doubtfully.

'Let's have a look, shall we?' he said, moving over to the ewe I'd brought in. He smeared a jelly-like substance over its side and took a scraping for testing. Bearing the sample, he disappeared out of the door in search of a microscope at the far end of the surgery. Like doctors, vets don't like to make quick judgments, but it still seemed an age before he returned, this time accompanied by a more senior colleague.

'Well, we can't find a scab-mite on that sample, but that doesn't mean that they're not there,' the junior vet said, looking, to me at least, a little disappointed. 'We'd better take another just to check.'

Together the two vets scraped even more vigorously over an even bigger area, before heading off to the microscopes once more.

This time it was close on half an hour before they returned, a look of mild puzzlement now working its way across both their faces.

'Well, there's still no scab-mite here, but I'm certain that's what's causing the trouble,' said the senior vet.

This didn't seem conclusive at all, so I repeated my reservations. Once more, however, they fell on deaf ears.

'There's a lot of it about this year,' the elder vet said, with a shake of his head. 'Best to treat the whole flock now, before lambing.'

We went through the treatment options for a minute or two, but there was really little to discuss. It was now too cold and too close to lambing to contemplate transporting the flock home to dip them. It would have to be the injection. So I ordered enough Avermectin to inject all two hundred and fifty of the Mortehoe flock.

'Should clear it up within a fortnight,' the junior vet assured me. 'There's never been any resistance to Avermectin.'

'It had better work,' I muttered to myself as I wrote out a sizeable cheque and drove home.

'You wouldn't have a little more of your chutney, would you, Debbie? This pot's running a little low,' Victor said, as he ladled a huge pile of the pickle on to his plate of bread and cheese.

Debbie shot me a sideways glance. 'Actually, Victor, it's only just been made. I'm not sure if it's ready yet,' she said, reluctantly producing a fresh jar from the cupboard. The attempt to preserve the supplies in the larder was in vain. The second jar too soon disappeared from view.

His impact on our food supply aside, it was a pleasant surprise to see Victor, a friend from South Devon. Like me he was a regular competitor at sheepdog trials, but – as with so many of my friends and colleagues – I hadn't seen him during the foot and mouth outbreak.

Victor was not one to stand on ceremony, so, as he'd been passing close by, he'd dropped in for lunch. I made the tea and we spent half an hour catching up on each other's news. The epidemic had missed Victor by a good many miles, but ironically it was only in the autumn afterwards that he had felt the worst of its effects.

During the latter part of the year, he usually put several hundred or more sheep out 'to keep' on surrounding dairy farms. Once the cows have been housed, it's common practice to have a flock of sheep on the farm to strip off any late autumn grass growth. This time, however, the bureaucracy had proved too much. The endless form filling, blood-testing, and the risk, albeit distant, of sheep bringing dormant foot and mouth on to their farms had persuaded many farmers that it was all too much hassle.

So Victor, having lost most of his usual winter's keep, had bitten the bullet and sold off a sizeable part of his flock. He was now supplementing his farming by making deliveries for a local pet food supplier. He didn't complain; indeed, it was a long time before he even mentioned his new career to me. Instead, he focused on the positive aspects of his life, and in particular on his one real love – his dogs.

It hadn't been long after he'd arrived that he let slip the news that, I suspect, he was keenest of all to pass on.

'I've got a little pup in the back of the Land Rover; you could have a look if you like,' he'd said over lunch. He knew I wouldn't be able to resist, so now, with his appetite finally sated and the last mug of tea drained, he and I headed out to the yard to view his new arrival.

Victor was soon leading out a rather large, gangling

eight-month-old puppy called Lad. He was tall and powerfully built, his head was broad, and his eyes had a calm wisdom that immediately put me in mind of Greg.

With a great bounding stride, the young dog was soon casting wide around the sheep in the paddock by the house. It was obvious straightaway that he had ability well beyond his age. Rather than running guided missile style at the sheep in front of him, as many young dogs his age would have done, Lad was watching his sheep and made a perfect outrun to the far side. Once there, he tucked his head down and brought them back the hundred yards, under near perfect control. Victor tried hard not to smile, but I could see beneath his great bushy sideburns that he was glowing with pride.

As Lad continued his display, he performed a whole series of tasks worthy of a dog twice his age, and when Victor eventually called him from the sheep, in itself a tricky manoeuvre with a keen young puppy, he trotted back immediately before positioning himself at his master's feet.

'Does he get his own dinner and put himself to bed in the evening?' I asked with an air of disbelief.

Victor could hold back his smile no longer and he chuckled for a second or two. 'Not displeased with him at all at the moment,' he said, rubbing the talented pup's nose.

'I could do with one like that,' I said, trying to contain my jealousy. 'Greg's only got another year of hard work in him. Lad'd be perfect as a replacement. He's got something about him that's just like Greg.'

Victor had picked Lad up at ten weeks old from Derek, a mutual friend of ours in Cumbria. Derek's reputation for breeding good working sheepdogs extended beyond not just the farming community but these shores. His dogs had been exported all over the world. It had been from him that both

Ernie and Fern had derived. Lad was the latest progeny of Derek's present stud dog, Jim.

'He's got another litter from the same parents,' Victor said, spotting my obvious interest.

Slowly the jealousy was being replaced by a different feeling: that this visit from Victor and Lad had been one of fate. It wasn't long into the evening before I picked up the phone to ring Derek.

———

I could feel my heart sinking into the mud of the Town Farm fields.

At first, as I'd delivered a fresh bale of silage to one group of sheep, I'd tried to ignore the fact that two ewes were rubbing themselves against the cold metal frame of the feeder. 'Can't be,' I'd said to myself.

In the three weeks since the sheep had been treated for sheep scab I had brought the flock back across the road, to start their winter feeding. Allowing for the life-cycle of the mite itself, by now there should have been a complete end to the problem, but as I climbed out of the tractor and started to take a look at the flock there was immediate evidence to the contrary.

First I saw one ewe scratching, then another. When I examined a group of half a dozen more closely, each of them bore the unmistakable signs of recent 'nabbing'. As if to offer final, incontrovertible proof, a sheep at the edge of the flock turned and chewed at her side, then turned back towards me with the telltale sign of a tuft of wool hanging from her mouth.

I felt a mixture of frustration and anger. The treatment I'd used was the best that there was; it should have been 100 per cent successful – but looking around it was clear there was no change. The same sheep were no better and no worse than they had been.

'I don't believe it,' I said to myself, sounding just like Victor Meldrew as I walked back to the tractor.

———

For once, the familiar drive through the glorious Cumbrian landscape was tinged with sadness. The rugged grandeur of the peaks and lakes was as impressive as ever, but there was a striking contrast between the hills today and when last I had visited. The foot and mouth outbreak had hit Cumbria as hard as any region of the country. It had been ruthless in its ravaging of the region's farming community. A year ago the fells and fields were full of sheep. Now the sight of a small flock was unusual enough to make you stop and look. The transformation was quietly shocking.

I had set off on the long drive up to Derek's farm at Lonscale, with two goals in mind. As I'd expected, convincing the children that we needed a new kennel member to replace Greg was easy. Even Debbie had been a comparative pushover. She appreciated how much I got out of training sheepdogs, and I think she also appreciated that a new dog represented a new start: a small but significant step into the future we'd talked about in recent days and weeks. Her only caveat was that the new dog must be 'a nice fluffy one, that looks like Gail'.

Derek had earmarked a member of a two-month-old litter for me. Perhaps as importantly, he had also agreed to let me mate Gail with one of his stud dogs, Ben. Since Colin had made his visit to the farm in the early days of the year, I'd received a steady trickle of enquiries about sheepdog puppies, all from farmers within the area. This in itself had come as something of a surprise. A lot of Devon farmers were very set in their ways, to put it mildly, when it came to their sheepdogs. Some didn't value them at all, regarding them as little more than a necessary nuisance. Others steadfastly refused to accept anyone else could be more of an expert on the subject than they were, and stuck to breeding their own dogs. Many worked on the principle that if they bred the old farm bitch often enough they'd eventually produce a dog that reminded them of the one good dog they'd had twenty years earlier. It was a method of breeding that had rather limited success. The fact that some were now prepared to buy in a sheepdog, and to do so from a non-Devon farmer, was a radical departure.

I knew I'd not make fortunes. The local farmers' opinion on the value of a sheepdog was summed up by a phrase I'd heard often: 'Never paid more than twenty pounds for a pup . . . always had good ones, mind'. However, the fact that I had an order book for half a dozen puppies seemed like real progress. So, as Gail had come into season, I had decided to take the opportunity and mate her with the prodigious Ben.

Ben had been the sheepdog world's most prolific stud dog in recent times, mainly because – as in every other area of life – nothing breeds success like success. So much of a sheepdog's potential is in the breeding, and Ben had produced a multitude of talented pups, including my own Fern. I had seen him in action on more than one occasion during visits to Derek and at various sheepdog trials. He was a wonderfully talented dog. Gail had her own qualities, and Ben's sharpness of movement was just what was needed to complement them. It seemed a promising match that had every chance of producing some first-class puppies.

It was dark by the time I made my way up the winding lanes through the woods above Keswick, and along the rough track to the farm nestling under Lonscale Fell. Like most of the farms that I had passed during the final hour of the journey, Derek's business had lost all of its stock in the epidemic. Few farms had as yet restocked but, with admirable determination, Derek had battled back. He had cleaned his farm and already restocked.

Despite the lateness of the day, he wasted no time in taking me out to a kennel on the far side of his yard. It had been fortunate that I'd booked one of the pups over the phone a couple of weeks earlier because the rest of the litter had been sold. My pup would have a long journey back to the south-west, but his siblings were travelling even further afield. One of the litter was bound for Japan, where there is apparently an up-and-coming interest in the working sheepdog.

Peering through the door, I saw a slumbering heap of black and white fur up against the far side of the run. A call of 'Here, pups' soon changed that. In an instant the five podgy-looking pups spotted the human company and ran over, whimpering and yelping, to stand on their back legs scratching at the mesh.

Derek reached in and lifted one out.

'Here's yours,' he said. I looked at it and could feel my face dropping. The pup that he held had almost no black on it at all, except for a splodge over either eye, and a spot over its back end that ran halfway down its tail.

A working sheepdog's appearance bears no relation to its ability as a working dog. It could be bright pink and still be able to drive sheep. It is no coincidence that when working dogs are registered with the International Sheepdog Society, there is no mention of either colouring or appearance.

However, for a shepherd, there are sound practical reasons for wanting a collie to have the traditional amount of black on its coat, not least the issue of familiarity. Once sheep learn to accept the authority of a dog, they will usually treat any new one of similar appearance with similar respect. A white dog, however, can become an object of curiosity for sheep, and lead to difficulties when it tries to work them.

A sheep's powers of observation are not necessarily linked to a clear thought process, however: I once watched a ewe chase a black lamb aggressively around a pen at lambing time, in the belief that it was black dog. Nevertheless, the conventional wisdom among shepherds held that, because of this, a white collie has to be a stronger working dog than its dark-haired counterpart.

All of which left me with a quandary, as I looked Derek's decidedly white puppy up and down. This was a little like choosing a new sofa: the colour wouldn't affect the comfort, but you'd still have to look at it for twelve years. The pup certainly didn't have a colour scheme that I would have chosen.

I tried to be diplomatic.

'He's quite, er, white,' I said with as much tact as I could muster.

'Do you not like him?' said Derek, just a hint of offence in his voice.

'Well he . . . it's just . . . he'll show up the dirt,' I stuttered, summoning up the only response I could think of.

For a moment I feared I'd jeopardised a good friendship, but the glint that suddenly appeared in Derek's eye quickly told me that he'd been having a little fun at my expense.

'Just as well he's going to someone else then,' he said, putting the pup back into the run and pulling out a second from the back of the pen.

This one had more black on it, though still certainly didn't carry the markings of a traditional collie. There was black on his face and over his back legs, but much of his middle was white except for a large black 'saddle' that sat over one side of his midriff. On the other side, the black stopped not far from his backbone, and a collection of black splodges covered the rest.

I was much more pleased with him than the other one. 'He's different,' I said. 'Once seen he won't be forgotten!'

As I gave the pup a closer inspection, I saw that his face was of the same dimensions as the brother that I had seen with Victor a few weeks earlier. He had the same broad head and strong jaw, with large overhanging top lip, all set around a large-boned frame. His eyes too were of the same type. Much is made of the eye of a sheepdog. Too light and the dog is often fixated by sheep and will tend to 'stick'. Such dogs will often lie down and refuse to get up and move, making them all but useless for work. Too dark and the dog may well be too weak, always turning away from confrontation.

'He'll be fine,' I decided after a few minutes. 'Not sure he's quite what Debbie envisaged though,' I said with a smile as we headed back towards the house.

As usual, I was shown the most wonderful hospitality by Derek and Helen. We sat down for dinner that night and were soon deep in conversation about sheepdogs, trialling and the more serious concerns of the farming industry.

Over the past few years Derek had begun successfully diversifying his business. As well as breeding dogs, he provided training courses to sheepdog handlers from all over the world, and on previous occasions I'd arrived at Lonscale to discover the house full of guests from near and far. In recent years, he and Helen had also begun staging working sheepdog displays during the summer months. Held in the fields at the bottom of the farm, they had been a considerable success and a useful money-spinner.

'We get a hundred and fifty people on a good night. It makes a lovely evening, at the foot of the fells watching the dogs work. You could do something like that where you are,' Derek said, after we had discussed the merits of the venture at some length.

At first I laughed. I certainly couldn't see myself 'performing' in public with my dogs.

'I don't think my dog skills are up to that sort of thing.'

'No, you'd be all right; your dogs are plenty good enough to show to people,' said Derek. 'You didn't do so bad in the English National. A lot of people would turn up just to see the pair of them working together, and you've got the tourist trade too.'

I could see he was right. Just as Lonscale was only a few miles from the thriving tourist gateway of Keswick, so Borough Farm was near Woolacombe, a seaside town that attracted visitors by the thousands during the summer months. An hour or so on a farm at the end of a summer's day would surely appeal to a few families.

'Hmmm. Something to think about,' I said, as Helen cleared away the plates.

It certainly proved to be that. With Gail successfully mated with Ben, I said my goodbyes and set off from Cumbria the following morning. On the long drive back to the southwest the idea of holding displays at Borough kept going round and round in my head. There were pros and there were cons; perhaps too many of the latter, I kept thinking.

On the positive side of things, Derek and Helen were right about our location. Set just off a main road, Borough Farm was ideally placed to catch the passing holiday trade on its way into Woolacombe. Windcutter Down, a chain of fields near the entrance to the farm where I often put the dogs through their trialling paces, would make an ideal spot to run displays. With a view across the valley and into fields half a mile away, it was both a natural amphitheatre and a well laid out trials course. My dogs too were more than capable of performing in front of a crowd. Greg and Swift had done well as a pair, or brace, at both the English National and the subsequent International Sheepdog Competition during the last season before foot and mouth shut the trialling world down. Fern too had become a creditable performer.

All the way down the motorway my mind swung from thinking it was a great idea one moment to thinking it was a terrible one the next.

Of one thing I was in no doubt, however. If we could somehow attract audiences anything like the size of the ones that Derek had in Keswick, the financial benefits were obvious. Mentally, I kept doing and redoing the sums. A hundred and fifty paying visitors for each display. A display once a week, for eight weeks. That, together with the sale of a litter of pups, and the dogs would be contributing as much to the farm's finances as the entire flock of sheep. By the time I reached Borough Farm, I'd given myself a headache thinking about it all.

Daylight had long since disappeared when the headlights lit up the side of the house and I pulled up in the drive. No sooner had I switched off the engine than I saw the kitchen door flying open and the children racing out. The arrival of a new dog was always a moment of high excitement and they had been allowed to stay up late to meet the new member of the household. As they ran out of the house Debbie followed behind, trying hard to disguise the fact – I suspect – that she was almost as excited herself.

I didn't open the rear door straight away.

'He's not quite your traditional-looking collie,' I warned Debbie.

'What do you mean?' she said, peering impatiently into the back of the car.

'Best to show you, I suppose,' I said. I opened the boot and took the pup from his cage. Almost immediately I was relieved of the bundle by Clare, who cooed at him for a moment or two before setting him on the ground to stretch his legs. Debbie looked first at the dog then at me in wide-eyed disbelief.

'He looks nothing like a collie!'

CHAPTER NINE
Beer and Skittles

Pubs aren't my natural environment, so as I stepped into the warmth of the Mariners Arms it was nice to see half a dozen familiar faces gathered around the skittle alley.

'Now here's a rare sight,' one of them said.

When I'd first got the invitation from the local branch of the farmer's union to join them for a skittles evening I'd been quite keen on the idea. Opportunities for farmers to get together had been few and far between over the previous twelve months. Both my normal means of socialising – livestock markets and sheepdog trials – had yet to be returned to the farming calendar, and, with the exception of a few visits from neighbours and friends, I had spent the year largely isolated.

The return of the mystery itching among the Town Farm flock had been niggling away at the back of my mind, and that morning I'd told Debbie I'd probably give the night a miss.

'Why?' she said, looking surprised. 'I thought you'd been looking forward to an evening out.'

I just grumbled – probably rather pathetically – about not feeling much like it at the moment.

'Go on, it'll be a good chance to catch up with people,' she said, waving away my rather weak case.

As I joined the throng of farmers already gathered, I was glad of her intervention. There were several familiar and friendly faces and everyone was indulging in a little light-hearted banter, with the skittle alley at the centre of attention.

Skittles is something of a Devon speciality. It's a traditional local game, similar to ten pin bowling, but with a shorter alley, and without any of the electronic sophistication. Anyone under the impression that it isn't taken as seriously as the more hi-tech version of the game would have revised their opinion pretty quickly if they'd watched the competition under way that evening.

There are a multitude of skittles leagues that run locally throughout the winter, and places within the teams are hard fought. Watching some of the balls being rolled down the alley this evening, it wasn't difficult to spot the regular players.

I took my turn to stand the skittles up for a game or two. I hadn't played more than twice in my life, so thought it might be a good way of picking up some pointers. I watched a lady in a long tweed skirt deliver a spinning ball that duly dispatched all the skittles on the alley. Then, twice in succession she sent the wooden ball down the alley with such force she again knocked down all ten skittles. Even on her fourth turn, when she left one single pin to knock down with her spare ball, she hit the skittle so hard it flew through the air, nearly removing my knee in the process.

'Sorry,' she called with half a smile on her face as she left the alley.

Faced with such opposition, I didn't expect to fare too well when my turn came to have a go. I wasn't wrong. While I did succeed in hitting the skittles on most occasions, I failed to get all ten pins down at any one time, and found that a single skittle left behind could be remarkably stubborn.

Having rather shown myself up, I decided to join some of those farmers who'd chosen to concentrate on the beer element of the evening rather than the skittles, and I soon found myself in conversation with a good friend, Richard, who had a beef and sheep farm a few miles away from mine.

In my experience, most farmers fit into two categories – hopeless pessimists or blind optimists. They either talk perpetual doom and gloom, or believe that everything will turn out all right, just so long as they work every hour of the day or night. Richard, on the other hand, was a rare breed. He was a born optimist, with a healthy sprinkling of realism. I'd always enjoyed having a yarn with him, listening to his laid-back approach to the problems which farming, and in particular sheep farming, throw up over the winter. Within a quarter of an hour or so we'd worked our way through the stories of endless mud, and inevitably headed towards the lambing season that was creeping up fast.

'Ewes wintered well?' I asked.

'They're looking well at the moment,' replied Richard. 'One or two slipped lamb over the last week, but I'm not worried about that, you always get it just before lambing.'

'Slipped lamb' is a local term meaning that a sheep aborts in the run up to lambing. Although 2 or 3 per cent of the flock will naturally lose their lambs in this way, there are also a variety of highly infectious agents that can sweep or 'storm' through a flock, causing massive losses to the expected lamb crop, so it can be a major concern.

As Richard and I had been talking, another figure had loomed into view, placing himself within easy earshot of our conversation. I cringed inwardly when I saw who it was. Although I didn't know him well, I knew of him. Norman kept a rather untidy smallholding not far away, but, as far as I knew, his real job was working on the roads for a firm of civil engineers. I also knew something of his reputation – for talking rubbish.

For a while he'd stood there, studying his pint. Obviously he had been listening to our conversation, and now he sidled over to join us.

'Ah,' he said in a high-pitched Devon tone. 'Slip lamb see. Always get it jus' 'fore lambin'.'

He was clearly going to offer us the benefit of his knowledge, so Richard and I exchanged a look and let him go on. 'I've 'ad it a time or two, and I've 'ad it out with them vets see, 'cos them don't know what causes it, but them won't admit it! But I know.'

I cast Richard a sideways glance, with half a smile. The causes of abortion were well known, particularly to Richard whose wife was a vet.

'What's that then?' I asked, unable to bear the tension much longer.

'It's in the rams see, stands to reason, you'm always get dead lambs at the beginning of the season, 'cos them rams ain't worked for the year. Them's using up the last year's sperm. Dead sperm, dead lambs. Like I say, stands to reason.'

Richard turned abruptly away, burying his face into his beer glass as he attempted to avoid breaking into fits of laughter.

I stood looking at Norman, biting the inside of my cheeks as hard as I could. Fortunately Norman was oblivious to other people's view of him, and was soon off into an endless list of sheep ailments for which he knew the cause, or the cure, or both.

Over the course of the next ten minutes or so we were treated to the tale of the sheep he'd brought back from the edge of death by feeding it with ivy, and the story of the stomach worms he'd cured with a mixture of pipe tobacco and copper sulphate. Each of them had achieved results beyond those yet managed by conventional medical science, mainly, it seemed, because each of them was mixed with linseed oil. 'T'was all my father ever used on his sheep,' Norman said, rocking away on the balls of his feet as he spoke. 'And 'e hardly had a sick sheep in all 'is life.'

In the face of all this, Richard and I were struck temporarily speechless, but Norman made sure the silence didn't go on for long.

'Itchin', that's another one that's a problem this time of the year. I had sheep scratchin' themselves maise the other year,' he said, reverting to an old colloquialism.

Up until this point I'd descended into a sort of bleary haze. I'd simply nodded, tutted and agreed at the appropriate moments in Norman's self-indulgent anecdotes. Now, all of a sudden my ears pricked up. Was he about to utter a word of sense for a change?

'I thought t'was the scab, so I put a sack over them 'n' soaked it in diesel, but after two weeks it still ain't got no better, so couldn't have been scab, could it?' he said, looking at me for an acknowledgment that his cure was completely foolproof.

'Well, what did you do?' I asked, clinging to the hope that

there still might be a molecule of worthwhile information to come from the conversation.

Norman had had to admit defeat on this one and had reverted to more orthodox treatments, which obviously hurt his pride. For a moment the sparkle in his eyes was lost.

'Turned out t'was fleas or lice or summit in the end; had to give them a dose of that stuff you pour on,' he said, referring to a standard treatment used for blowfly. 'Kills lice as well see. Did the job in a couple of weeks.' For a moment I carried on nodding absent-mindedly, but suddenly one word came leaping out at me. 'Lice,' I said inwardly. 'Of *course*, lice, why the heck didn't I think of that?' It made perfect sense. Lice, while an unpleasant problem, are nowhere near as threatening to the welfare of a flock as scab. They aren't as contagious either, which would explain why they hadn't spread beyond a few members of the flock.

As my mind worked overtime, Norman remained silent, looking a little solemn for a good thirty seconds. The peace wasn't to last, unfortunately, and soon he'd brightened up, ready to share one final pearl of wisdom with us.

'Mind you, I reckon it would've cleared up even quicker if I'd got it mixed with some linseed . . .'

With his pint glass drained, Norman was soon sloping off back to the bar, leaving us to return to our conversation.

'Well, what's new on the farm with you?' Richard asked. 'Taken on any more ground lately?'

'No, and I'm not sure that keeping more sheep is the answer anyway, the way things are. By the time you pay the rent on the ground, you end up working for nothing,' I said with an air of resignation.

'Perhaps you'd better get Debbie to start doing bed and breakfast,' Richard said with a grin.

I gave him a half-hearted smile. Bed and breakfast had

always been the most obvious source of alternative income for farmers in this part of the world, particularly given its popularity with 'visitors'. In recent years, however, local farmers had become involved in far more ambitious schemes, the most prominent being campsites and golf courses, but perhaps the most remarkable success had been achieved by two local men, one a former sheep farmer, the other a dairyman, who had turned their holdings into farm 'theme parks'. Both had proved to be hugely popular and annually attracted thousands with their demonstrations of shearing and milking, lamb-feeding and falconry. They'd brought many much needed jobs to the area in the process.

'They reckon more than half the country's farmers have got a second business now,' Richard said. 'What's yours? Unless you've got some magic formula for making money out of sheep that you're not telling us about.'

'Oh I have, the key is lacing my rams' water with linseed oil before the tupping season,' I said, nodding towards Norman, who had limpeted himself on to another pair of unfortunates near the bar. 'Increases their fertility by 1,000 per cent.'

Richard laughed, and then quickly ducked his head as Norman looked once more in our direction.

'No, it's not the sheep that I've been thinking about, it's the dogs.'

I told him about my visit to Derek in Cumbria and the idea he'd planted.

'But it's hard to know how many people would come to watch a sheepdog display on a warm evening when they could be on the beach instead,' I said, articulating one of my biggest recurring doubts.

'Interesting idea.' Richard nodded before taking another long, slow sip of his pint. 'What would you need to set it up, do you think?'

'Not a lot, to be honest,' I said. I explained to him my idea of holding the demonstrations on Windcutter Down. 'All I'd need to do is stick a few bales of hay out for people to sit on. The main job would be advertising it.'

'Well, for what it's worth, David, I think you'd get a decent response,' he said eventually. 'People come down here for a taste of the country life. They'd love to see your dogs at work, I'm sure, especially the children.'

'You think so?'

'Yes, definitely – and besides, what would you have to lose? It's not as if it would cost you anything much, a few posters to be printed and that's it. It's something you'll be doing with the dogs anyway, when the trialling season gets going again,' he said, draining his glass.

'True,' I nodded, realising I hadn't thought of it that way.

'So that's settled. I'll expect to see the posters in the Woolacombe Tourist Information Office in June,' Richard said, picking himself up from the chair, ready to head back to the bar. 'In fact, why don't we drink to your new venture now. Another one?'

———

In the wake of my chance conversation with Norman I looked up the section on lice in my copy of the sheep farmer's bible, the *Veterinary Book for Sheep Farmers*. The subject wasn't covered in any great depth, but there was enough there for me to reach a couple of fairly firm conclusions. First, lice are nothing like as infectious as scab-mites, which explained the sporadic nature of the outbreak I'd seen in the Mortehoe ewes over the last six weeks. Secondly, lice were not skin-feeding parasites, and therefore wouldn't have been susceptible to the highly expensive treatment I'd bought from the vets.

Loath as I was to admit it, Norman had been right about the

cure as well. A quick conversation with a sales rep from a farming supply company revealed that one of the new 'pour-on' treatments, highly effective in preventing blowfly strike during the summer, also killed lice. The cost was a fifth of the price of the injections I'd given and was a lot easier to administer too. I felt relieved, and hopeful that I'd now be able to sort this problem out. Yet a part of me was still quietly seething at the vets' inability – or unwillingness – to consider anything other than the obvious answer.

The treatment needed to be administered as soon as possible, so the first dry day I headed for Morte Point to round up the infected flock.

When March came, they would need to come back to the more sheltered ground of Town Farm to start the winter feeding of silage and 'cake' in earnest, but for now, with their nutritional needs still low, I was happy for the flock to forage what they could among the crumpled brown bracken and ochre grasses of the headland. The only drawback with this was that, as pickings became more and more scarce, so the flock had become dispersed, rooting around on the remote outcrops and ledges. With this in mind, I had – for the first time in weeks – brought a full complement of dogs out to work. There was no longer any real reason for Swift to be indoors. The wound on her foot was now fully healed and she was clearly feeling no ill effects from the loss of the toe. Better still, the vet had rung to confirm that the lump had been benign. Inactivity doesn't do a dog any good, so I'd decided it was time she got back into the swing of things. With Gail and Greg as alternate partners for Swift, and Fern and even Ernie in reserve, I had more than enough dog power.

It was a typical North Devon late winter morning. The dull, lifeless sky seemed to be made up of a single mass of clouds that was carrying a chilling dampness up from the sea.

The gather started and the first group of thirty or more ewes turned inland with little resistance. Gail was once more working well and it was dawning on me that I was beginning to rely on her more than Greg. He, for once, was tucked in against my leg, watching as his daughter took on a role that for many years had been his and his alone. He looked towards her, and then up to me. I'd had given more than a penny for his thoughts.

A few small packets of sheep were grazing in the least accessible corners of the Point. Half a dozen were visible among the thickest cover, not far from the old coastguard lookout. This holds a special place in the history of Morte Point. It carries a commemorative plaque to the men of the village who kept watch from a small shelter throughout the years of the First World War. The lookout was only removed in the 1970s, a deeply unpopular move that, according to many people in the village, was carried out surreptitiously under the cover of a heavy sea mist.

The sound of my whistle and the sight of other members of the flock in motion far below was enough to get these ewes moving, but another group couldn't find it in themselves to be as co-operative. Morte Point is dotted with landmarks that have over centuries acquired their own, unique names. On the north side, near where I now stood, there is a spectacular inlet know as 'Wide Gut', one of four guts along this stretch of coast. 'Wide Gut' drops to one hundred and eighty feet deep. On the western edge of this, overlooked by a steep cliff, is a triangular promontory called Morte Wells Island. It's not strictly an island, as a causeway of perhaps fifteen feet wide joins it to the mainland. During the summer the sheep can rarely find their way out there, but, as the overgrowth dies away, the path becomes accessible once more.

Six ewes stood looking up towards me from the lowest edge

of the island. They showed no intention of joining the rest of the flock.

This is the only run which I avoid asking the dogs to make, if at all possible. For them to circle the sheep, they must go along the edge of the cliff, picking their way from rock to rock. From my viewpoint, it looks as if they are sprinting along the very edge of the precipitous drop. I find it a nerve-racking spectacle.

There was no option today, however, and I chose Swift to make the run. She is the most controllable of the team and had done it many times before. She knew where she was going and a few seconds after having dipped out of sight down the steep path in front of me she re-appeared on the causeway. I blew a slowing whistle as she started along the most dangerous section. She eased up and picked her way across, only once turning towards the edge in a way that made my heart flutter.

No sooner had she run across the narrow approach than the sheep spotted her. With a dog now blocking their one escape route, they opted to evade capture by dropping down on to the rocks on the furthest edge of the island, disappearing from my line of sight as they did so.

Swift had seen them, but paused as if awaiting instructions as to how to proceed. Below her was one tiny gully, only a few yards wide, but plenty deep enough to cause a disaster, and, looking across the landscape, it would be easy for me to underestimate the hazards invisible from this distance. I took a moment to weigh up the situation. The sheep simply had to be gathered. Neglecting to treat a couple would render this whole operation useless, as the untreated ewes would simply re-infect the rest.

Asking any of the young dogs to go on from here would have been out of the question, but such was the level of trust I had in Swift that I felt we could carry on. I blew a left-hand whistle as gently as I could while still being audible. Far below me Swift turned her head and glided off to her left. In a second she too had jumped down and out of sight.

Experience has taught me that these moments, however short, seem to last for ever. So I looked at my watch and decided that if there were no sign of sheep or dog within one minute, then I would recall her and make the walk myself.

I needn't have worried. A sheep appeared over the rocks, turned defiantly for an instant, then jumped up on to the grassy heathland. Straightaway she was joined by a second, then a third, until a few moments later Swift appeared behind. I counted six ewes, the correct number. Soon they were heading in the right direction, and Swift was at my side.

'Good to have you back at work,' I said, briefly rubbing the nose of the consummate professional.

———

I arrived home to find Debbie and the children at the small pond, once more trying to chase the ducks into their run before nightfall. This time, however, their rounding up operation was being hampered by the presence of the new puppy. While Debbie and the children attempted to drive and coax the ducks from the water, he was darting to and fro in the field behind, indulging himself in one of his favourite pastimes, diving on molehills. The dog seemed to have trebled in size. He now stood nearly knee high, and the enormous length of his legs, coupled with his unusual markings, only served to underline people's scepticism about his breeding. 'Has he a little bit of collie in him?' I had been asked on more than one occasion by visitors, puzzled by his appearance.

His eccentric looks – not to mention his behaviour – had endeared him to Clare in particular, but this evening his efforts to unearth moles by first flattening each molehill then flicking the loose earth into the air with his nose, was having the effect of chasing the ducks back on to the pond.

'Jake, stop it,' I heard her shouting as I arrived to join them.

The gangling white pup had taken root in the kitchen since my return from Cumbria. My protestations that he was quite used to living outside had fallen on deaf ears. He now spent much of the day snoozing in a cardboard box, which the children had moved next to the Aga, and the evening chewing on a table-leg or one of the children's socks, which he found much more appealing if it had toes inside. No sooner had we finished rounding up the ducks than he was in the kitchen, indulging in the latter again.

'Ow, Jake!' Laura yelled as the pup yanked repeatedly at her foot, dragging her hopping across the kitchen floor as he did so.

'Take your toe out of that poor pup's mouth, he doesn't know where you've been,' I teased Laura, ignoring the pained expression on her face. 'And anyway, what's this about Jake?'

'Nick has been reading a book about a dog called Jake, so that's what we've decided to call him,' Laura explained. 'The dog in Nick's book is really silly though,' she added as an afterthought. There are a few traditional names, such as Moss, Spot and Ben, but in recent years the trend has been to give male sheepdogs male names, as if they were an everyday workmate, which of course they are. Names like Davey, Jim, Craig, Pete, even my own Greg and Ernie, have become popular. It is, ultimately, a matter of personal choice, but there is one rule of thumb. To ensure a shepherd can issue quick, clear instructions, a dog's name should always be short, preferably one syllable. On that basis 'Jake' seemed perfectly fine to me.

'He knows his name already, Dad. Watch this,' said Laura,

who had now been joined by Nick. Soon the pair of them were shrieking together: 'Jake, Jake, Jake.'

In fairness their calls did momentarily attract the pup's attention. He sat on the floor, looking a little bemused, but then he stood up, squatted slightly and deposited a small puddle on the tiles.

'Looks like you've taught him his first command.' I smiled at Laura. 'All you need to do now is work out how to get him to use a mop and clear it up.'

CHAPTER TEN

An Early Start

I n the paddock above the house Clare was accompanied by
two figures, glowing a ghostly white in the thin morning
light. With admirable dedication, Clare would every morning
take Minty to the field, and dung out his stable, before going
to school. Recently, however, they'd been joined by Jake. This
morning, as usual, he was lolloping along behind, generally
behaving like the carefree – if slightly eccentric – young pup
that he was.

In the past few weeks Jake had begun displaying habits that
I'd never seen in a collie before. Each morning, he'd take great

pleasure in pouncing on the first object that he stumbled upon when he left the house. Often it would be a stick or a small branch, sometimes it would be a clump of grass that he would rip out of the ground with his teeth. He would then proceed to trot proudly ahead of Clare, much like a dalmatian. In fact, with his black-spotted coat he could well have been the result of some impromptu liaison between a dalmatian and a collie.

He had also developed a fascination for the hinges of the gate that led into the paddock. As Clare opened it each morning, he would prick his ears and stare intently at the hook of the hinge, following it with his head as it opened, and eventually shut. Then, with a baffling precision, he would dive on the bottom bar of the gate, biting as if to catch some unknown quarry.

Border collies that are bred for work have some common traits that are obvious from early on in life: an interest in herding sheep, ducks and chickens, is the obvious one. Jake, however, was a dog unlike any I'd ever seen before. I already had my doubts whether he was going to be a working dog.

The final part of Clare's morning routine was collecting the eggs laid by her small flock of hens. She'd first acquired some as a Christmas present a few years previously, but the original birds had fallen prey to the visiting fox within the first few months of their arrival, so the replacement birds were now enclosed in a large, wired pen that resembled a small prison. They were getting old, and had as a result starting laying fewer eggs. This morning, as I walked across the yard, Clare appeared in the drive looking a little dejected. I knew the answer to my question before I asked it.

'How many this morning?'

'None, again,' Clare said. 'Dad, what are you going to do about it? That's the second morning running with not a single egg, and we've only had three this week.'

'You need some younger birds, then they'll start laying again, now the daylight's a bit longer.'

'But what shall we do with the old ones?' She looked at me, but before I could answer she huffed, 'Can't we just release them in the woods to live the rest of their days as free range?'

She was soon heading off indoors, to get ready for the school bus that would be pulling up at the top of the lane in just twenty minutes. Back in the yard I let all five dogs out of their run to join Jake. As Greg, Swift, Gail, Fern and Ernie chased around, bursting with morning energy, Jake hesitated for a moment, looking up at the back door through which Clare had just passed. It was as if he was weighing up his loyalties.

He might be an odd dog, but his instincts were still canine. He was soon taking up his place within the pack. The process by which Jake had come to learn his rank was a mystery to me, but within a short time of arriving on the farm, he somehow knew that he belonged in the lower ranks of the pecking order, though curiously not right at the end.

He seemed to have settled his chosen place as being between Fern and Ernie, rather than after Ernie as would have seemed logical. The only possible explanation that I could guess was that Ernie was still so much in awe of Greg that he chose to keep the furthest possible distance from him, in order not to provoke any disciplinary action.

As ever, Gail didn't join in the hierarchical jostling. In fact, now in whelp, she was even keener to avoid the mêlée of the pack and a confrontation with Fern in particular.

With lambing just ten days away and most of the Borough Farm flock now housed, the first part of my morning was spent with the regular winter routines: strawing pens, dropping in bales of silage and feeding the high-protein 'nuts' that supplemented their diet during these crucial final days.

Such is the anticipation of the ewes at feeding time that as soon as I open the shed door they burst into a cacophony of bleating that almost hurts the ear. It was no different this morning, as I began pouring the nuts up and down the feeders to the four hundred ewes I had under cover. As I cast my eye over them, however, I was drawn to one corner of the shed and an unwanted surprise. In the third pen, among the deep straw, lay two feeble-looking, wet, white bundles: the first – very early – arrivals of the new season. A sheep's normal gestation period is between one hundred and forty-five and one hundred and fifty days. Lambs born before this time have a limited chance of survival. These two were closer to one hundred and thirty-five days and it was a wonder that they had drawn breath at all.

I climbed in over the rails to where the lambs lay, wet and shaking. When I picked them up in turn, both were cold to the touch, and when I pushed my finger into their mouths, it didn't need a thermometer to tell me that each was severely hypothermic. Worse still was their appearance. Hypothermic lambs can often be revived with the right treatment, but these two both had all the hallmarks of premature lambs – ears tucked down, a slightly bald appearance across the forehead, and a fleece that was yet to develop fully.

Common sense and experience told me that these lambs should be put down, but somehow, after all the troubles of the past twelve months, these signs of new life on the farm seemed too precious to let slip so easily. Fortunately, the modern shepherd has access to innovations that his ancestor could never have imagined. At lambing time, none is as useful as the lamb-warming box, a wooden container with a mesh floor, through which waves of warm air can be blown.

I took some hay from a nearby feeder, and rubbed the lambs in the hope of lifting the chilling wetness and to stimulate some movement. I then turned on the warming box, laid the two in its cosy depths, half closed the lid and left them for a while.

Under normal circumstances, a sheep that has recently produced will bleat relentlessly, looking for her lamb, but it is not unusual after such a premature birth for the ewe to show no such maternal instinct. So it had proved today. There was no obvious sign of the mother, so I got on with feeding the ewes first.

With the shed returned to peace and quiet, however, it did not take me long to locate a sheep with a blood-stained back end. Whether she liked the idea of motherhood or not, I needed her. A lamb needs its mother's first milk, colostrum, within the first hours of birth, as this contains the antibodies to fight off life's first challenges. The ewe didn't have much milk to offer, but I managed to take 80ml from her, enough for a first feed for two such small lambs. Neither would be strong enough to suck from a teat, so I unwrapped a stomach tube from the cupboard, and, sucking half the treacle-thick colostrum into the syringe, threaded the tube into the first lamb's stomach and gently squeezed.

It is a very simple process to stomach tube a lamb, and it saves countless lives each year. It also saves dozens of hours of work for the shepherd, who is spared having to teach reluctant patients to drink from a bottle. Once the lamb is stronger the urge to suck usually returns and it can be reunited with its mother. Although that seemed a distant prospect for this pair, I still wanted to give it a try.

Its stomach feed complete, the second lamb gave a half-hearted bleat as I replaced it in the warming box. I felt its mouth once more. Still stone cold. Warming a hypothermic lamb is a slow process – too much heat too soon can be as dangerous as too little – so I checked the temperature before leaving them. Only time would tell how these two would fare.

Outside there was an almost spring-like brightness to the day and the change in the weather seemed to have energised the farm's birdlife. A small flock of chaffinches were once more devouring the grain spilt around the yard. Among the brambles below the farm a charm of goldfinches were picking over the last of the winter seed heads. Elsewhere the thriving population of house sparrows had ventured out of their nests behind the weatherboarding of the sheep shed to criss-cross the yard in search of pickings.

Of all the farm's birds, the robin was the one I most associated with the darkest months of the year, but in the sheep shed even his song seemed to ring out with a renewed intensity, as if he too felt that winter's grip was finally weakening.

For now, however, daylight was still a precious commodity and the rest of the morning and early afternoon provided barely enough time to do the rounds of the farm. I spent an hour feeding and checking on the flock out at Mortehoe, then tried to fit in a little fencing. The work at the far side of the valley was still progressing at a snail's pace. Today, once again I managed to squeeze in little more than a couple of hours digging in posts.

By the time I returned to the farm the children were arriving back from school. Laura was the first to join me. Her uniform shed in exchange for something more work-like, she catapulted into the shed in search of the new arrivals.

'Mum said we've got some lambs,' she said, thrilled, not bothering with any preliminaries of conversation.

'Well, they're not much to get excited about, Laura,' I said, not wanting to dampen her enthusiasm too much, but not wanting to build it too much either. 'They're rather premature, and I'm not sure if they'll live, but they're in the hotbox if you want to have a look.'

Laura was there in a flash and opened the lid cautiously.

'Dad, it looks like this one is dead,' she was soon calling over her shoulder.

When I'd fed both lambs a second time, just after lunch, the smaller of the two had scarcely been able to hold up its head. Now it had succumbed to the inevitable. I reached into the box and took it away.

As I dealt with this one, Laura busied herself inspecting the other, which had just managed to lift its head.

'Can I feed him?' she asked.

'He's too weak to suck, but you can help me to stomach tube him if you like,' I told her.

I began the process of catching the ewe, milking her out and, with a little help from Laura, squeezing the contents of the tube into the lamb's stomach.

Laura held the lamb for a minute or two before putting him back into the box.

'He's so sweet, I'm going to call him Lucky,' she said.

'I don't think it's a good idea to start giving the lamb a name; he's only got a small chance of survival,' I cautioned.

Growing up on a farm, the children had learned the realities of life and death. The old adage 'Where there's livestock, there's dead stock' is not a flippant comment from a dispassionate farmer, but a countryman's acceptance of the way things are, and always will be. Laura, like her brother and sister, had grasped it from an early age.

'Well, if he dies, we can call him Dead Lucky,' she said, giving the lamb a final few strokes before heading back to the house for her tea.

Ten days after the arrival of the premature pair of newborns, the lambing season erupted. As usual the clock ceased to have any meaning as an endless torrent of life started to pour forth, at all hours of the day and night.

A 'dry time' is all that most shepherds will ask for over the lambing season. Newborn lambs will put up with icy winds and frosty nights, but add some wet into the equation and they find survival a much tougher battle.

Of course, I'm not nearly as vulnerable to a swing of the wind as my predecessors. Not so many years ago lambing was carried out exclusively in the open, with shepherds living in makeshift shelters among their stock. As a child I can remember being taken to a lambing field ankle deep in mud, where the most vulnerable lambs were sheltered under rickety constructions made of straw bales and corrugated tin. Yet, even with all my modern advantages, the sight of a TV weatherman announcing that this evening 'winter will show a nasty sting in its tail' was most unwelcome, especially as he was doing so with a glib smile that suggested he had no idea his news could wreak havoc for the thousands of sheep farmers across the country who were currently going through the most crucial phase of their working year. 'So if I was you, I'd stay inside by the fire for the next twenty-four hours,' he advised with a rather patronising smile.

'As if,' I growled at him.

Once more, alas, the lunchtime forecast was spot on. As the day wore on, the skies grew darker and more threatening. By mid afternoon the gentle rustle of the woodland on the bank below the shed had become a vicious thrashing. With the wind now swung from the west to the east, the gathering storm was cutting relentlessly in through the open side of the lambing shed. The new arrivals had to take priority, so I abandoned any outside work, leaving those lambs already in the fields to find what shelter they could against the walling and woods.

Inside the shed, the iciness of the wind was so severe that even the strongest of the newborn lambs were becoming cold before they could stand. I moved the latest deliveries quickly from the bigger ewe pens into the individual pens with their mothers. The deep straw and solid sides of the smaller pens would keep the chill at bay.

With evening now setting in, I didn't need to look outside to tell the weather was worsening by the hour. As darkness fell, a flurry of snow was whipped across the shed, settling on the backs of the sheep and dusting the straw. I was just moving the last of the newborn lambs, when things took an even less helpful turn. A sudden flash from outside filled the shed with a strange blue glow. The lights flickered, dimmed and briefly recovered before, with the second mighty flash, they failed completely. It had come from across the field some two hundred yards away. Squinting into the wind, I could see light dancing on the top of an electric pole. Below it, power cable lay on the ground.

Back at the house I discovered that all of Mortehoe had lost power. Debbie had already managed to phone the electricity board, but neither of us were hopeful they could resolve such a major problem in the middle of such a wild storm, so I headed back to the lambing sheds, where I began rigging up the generator to a pair of halogen lamps fixed to a stand. The twin beams left half of the shed in shadow but at least made work possible.

To my amazement the cavalry arrived within forty-five minutes. A convoy of white Land Rovers drew up in the yard and a team of men in yellow PVC emerged. The local engineers must have cursed Borough Farm. The last time they had been called to repair damaged wires had been on a particularly wild Christmas Day four years previously. Tonight's power failure seemed more serious to me, but I didn't have the time to speculate. Instead I just directed them across the field to where I'd seen the pole flashing. Within a few minutes the spot was illuminated by a ring of headlights and halogen lamps.

As the evening wore on the storm showed no sign of abating. Around nine o'clock Debbie appeared in the shed, clad in waterproofs and carrying a torch and a flask of coffee she'd taken down to the engineers, still hard at work in the field.

'Why don't you come inside to warm up for a minute?' she asked. 'Seems quiet here at the moment.'

The house was lit by a dozen assorted candles, and enough heat was coming from the Aga to warm the kitchen and boil the kettle. For an hour I stood against the stove, clutching a steaming mug of tea, grateful for a little comfort.

To my amazement, I returned to the shed to find the power back on. Somehow the engineers had been able to winch a new line into position, even in this weather. Quite a miracle, I thought to myself. I stepped out into the yard to see the white Land Rovers passing by, and gave them a grateful wave as they went.

The evening round passed off without incident, another dozen lambs arrived, each entering the world without any complications, and by just gone midnight I was able to slosh back across the yard, ready for a couple of hours in bed.

At least the flurries of snow had moderated to driving rain.

———

The lambing season is always energy sapping, and this year's was certainly running my batteries low. The first rounds of the morning had been starting long before daybreak and the last, like tonight, ended past midnight. I'd go to bed dreading the electronic ringing of the alarm. Every time it woke me, I felt more tired. The perpetual wind of today had left me even more worn out than usual. So, when the high-pitched buzzer went off at 3 a.m. – less than three hours after I'd drawn the duvet over my head – I was all but comatose. Outside, the rain was still battering on the bedroom window in waves, the house was cold and in darkness. With my eyes still shut, I sat on the edge of the bed and threw on the first articles of clothing that came to hand. Then, opening one eye for safety, I stumbled down the stairs and pulled on the stinking lambing overalls that lay by the back door.

Trudging sleepily across the yard, I hoped there would be the bare minimum to deal with, but, with thirty ewes having produced during the previous day and scores more yet to deliver, I knew there was only a slim chance of me being able to head immediately back to bed. And so it proved. There is a general belief among sheep farmers that bad weather always seems to increase the rate of births. Some argue that it's proof of the sheer cussedness of sheep, but I suspect there's a more scientific explanation. Whatever the reason for it, as I squinted under the powerful electric lights in the shed I found the place alive with activity once more.

A further eight ewes had produced since I'd left three hours earlier. I set about moving them to their individual pens with their lambs, but as I walked into the main pen to collect the third ewe with her lambs, the torchlight caught a dark shape, half hidden in the straw-filled shadows at the corner.

I bent down and picked up a large lamb, dead and cold. The cause of its demise was obvious at a glance. Its head was swollen grotesquely, due to strangulation during the birth process.

It's not uncommon. Lambs will often 'present' with both of their forelegs facing backwards, rather than in the normal position with the feet just in front of the emerging nose. Some ewes will still manage to deliver successfully, but if the lamb is big – as this one was – the shoulders will wedge behind the ewe's pelvis with often-fatal consequences.

Caught in time, it's a situation a shepherd can avert. It can be a difficult process, but it is usually possible to manipulate the lamb so that it can pass out into the world. In this instance I hadn't been there to help, something for which I immediately blamed myself. I felt as if I had let down both lamb and ewe.

Disconsolately, I picked up the lamb by its back legs. He was indeed a whopper. As I carried him across the pen, his bereaved mother gave a gentle murmuring bleat. Despite the turmoil of the last few hours, her maternal instincts were still intact. When I looked more closely at her, I saw her udder was large and brimming with milk. There was still one way to take something positive from the situation. I had several lambs belonging to ewes with insufficient milk, all of whom were waiting for a suitable foster mother.

Sheep are notoriously reluctant to rear offspring other than their own, but there are a few ways of circumventing this, many of them based on age-old techniques. With a heavy sigh, I took the knife from my pocket and gave it a sharpen, ready to try perhaps the oldest of these – skinning the dead lamb and placing its coat on another newborn.

Perching on an upturned bucket, I laid the lamb on its back and slit its skin from the belly upwards. As I did so a gust of wind cut through the shed; blood covered my hands and spread up my forearms; suddenly my eyes felt more tired than ever before; I felt my spirits sagging. In years gone by, I'd considered lambing a magical time, a time of rebirth and renewal in more than one sense. The sight of the first lambs of the year had always restored my faith in the work I was doing. Now, my head was filled with darker thoughts. Did I really need this job that I had to work so hard to preserve?

A few minutes later, with a lamb wrapped in the skinned coat of the other, the fostering was successfully complete. The ewe had accepted the scent from its dead offspring and allowed the youngster wrapped within its coat to suckle. I watched the fostered lamb feed for a while before finally turning out the light.

By the time I crawled back into bed it was approaching 4.30. I was freezing and exhausted, but my brain was fully awake. I reset the alarm for two and a half hours later. Perhaps I might be able to persuade Debbie to make the early morning round.

CHAPTER ELEVEN
In The Blood

Two weeks into March and the promising green shoots of early season grass had become shrivelled and tinged with purple. As the first batch of ewes and lambs scrambled their way out of the back of a trailer into the fields at the far side of Borough Farm there were meagre pickings for the mothers and their offspring. The ewes would have to be offered cake for a week or two yet – more expense when we least needed it.

As the ewes slowly began to locate their lambs, I worked with Gail and Swift to shed off the family groups and drive them to different parts of the field – with differing degrees of success. Like her father, Gail could tell from a few simple signals exactly which individual sheep she was being asked to shed. Working with her reminded me of the level of two-way communication that had made Greg such a special dog for me. Although Gail had a long way to go before she could match him for brains, she was now showing many of his best attributes. Unfortunately, however, it wasn't just the dogs who were demonstrating the attributes they had inherited from their ancestors.

A first time mother, or shearling ewe, Blue 238 was determined to show that she was descended from extremely cussed and stupid forebears. She was also showing no sign of having inherited any maternal instincts whatsoever. She had caught my eye back at the farm, where she had been particularly obstinate to load into the trailer. Now she seemed the keenest to get away from the rest of the flock – including her lambs. While the rest of the bunch stayed close by, making the occasional call to their offspring as I unloaded them from the front of the trailer, Blue 238 took herself further and further across the field.

After two decades of watching sheep in action, I'd begun to understand the way they think. In the main they conform to a standard set of behavioural patterns. The fact that this ewe had given neither a bleat nor a glance in the direction of her lambs was a pretty strong indication that she had decided that motherhood wasn't for her. She confirmed this when, rather than turning tail at the sight of Gail approaching her, and running to protect her lambs, she decided to make for a point at the other side of the field that marked the furthest possible distance from them.

This was a new situation for Gail to deal with, something that was far outside her current understanding of ovine behaviour. Blue 238 turned and twisted, determined to break past her, then, realising there was no way through, resorted to running straight at her, bouncing all four legs in the most aggressive stance that she could muster.

To her credit, Gail stood her ground, but there is little that the most determined sheepdog can do if a ewe decides literally to run over her, which was precisely what Blue 238 proceeded to do. For a brief moment Gail was on her back as the ewe headed into the far corner of the field.

I gave a word to Swift and she was soon on her way to assist her somewhat bemused daughter, already passing the errant ewe before Gail could regain her bearings. With the ewe's escape route blocked, Gail rejoined her mother and – side by side – they began to persuade the ewe back towards the rest of the flock. The others were by now anxiously working their way through the throng of bleating lambs, trying to identify their own.

I caught the two Blue 238 lambs, and carried them towards their mother. She was still resisting and had faced up to the dogs for the entire two hundred yards of her retreat to the main flock. As I pushed the two lambs forward, they bleated and ran to her, but there was no maternal reaction.

'That'll do, Gail; that'll do, Swift,' I said, releasing the two dogs from their duties momentarily. 'Let's see if the old cow is going to look after them now.'

Not a chance. Without so much as a sniff towards the lambs, she was off again – but this time the dogs were ready. She'd run no more than thirty yards before she was forced to walk backwards to me once more.

This couldn't continue, I decided. So after a brief battle using my crook, both Blue 238 and her lambs were back in the trailer. Perhaps a few more days living in close proximity to her offspring in a lambing pen would see her maternal instincts rekindled.

———

'The quicker they lamb, the sooner it's all over with.' It had been Debbie's sister, Judith, the Kentish shepherdess who had been my mentor when I was a young training-scheme apprentice, who had first imparted these words of wisdom. So it turned out this lambing season. After a little more than two hectic weeks the shed was half empty. Five hundred ewes had lambed successfully and the fields on either side of the valley were dotted with hungry sheep picking at the first fresh shoots of the season. In the majority of cases, each of these ewes was accompanied by two lambs, a fair indication that the season had gone well.

Only the Mortehoe flock, which had spent the entire winter outside, were still producing at any sort of speed. This morning, the last Saturday in March, the children had joined me on a visit to the lambing fields of Morte Point.

Part of me hoped Nick, Laura and Clare's eagerness to join me had something to do with their enthusiasm for shepherding – all three of them enjoyed elements of the work – but, today, I sensed it also had a lot to do with the weather. Spring had sprung suddenly and the hillsides surrounding Mortehoe were suddenly ablaze with glorious burning-yellow gorse blossom, so bright as to seem almost unnatural. The air too bore the unmistakable hallmarks of spring, fresh, warm and strong, laced with the gorse's coconut-sweet scent. Early lambing this year had been something of nightmare. In these conditions it was a joy.

I restrict the sheep that are yet to lamb to one field running alongside the lane. The grass here is soon eaten off tight, but I provide a constant supply of silage and enough high-protein sheep nuts to fill any gap in their diet. As we arrived this morning Nick and Clare spotted a newborn lamb at the higher end and set off to inspect it. Laura and I headed in the other direction and had soon counted four ewes with lambs that had arrived overnight.

The number of lambs is not my greatest concern with the Mortehoe flock. The Romneys here are kept on a low input and low output basis, so the fact that only one ewe had a pair of lambs and the other three all had singles mattered little. The most important aspect was that the lambs were up and fed, which they were. Even better, all five lambs were tottering along behind their respective ewes. All in all the scene looked a picture.

Clare, in particular, had developed an excellent shepherding sense.

'There's one ewe with a pair, and one with a single,' she called as she returned from her foray to the top of the field. 'One of the lambs had wandered off, so we put it back with a ewe. I think she likes it.'

I didn't doubt she was right. Spotting the telltale signs that indicate a ewe is happy with its lamb was almost second nature to the children now. In this respect, shepherding was already in their blood.

My lambing system for the outdoor sheep is as simple as I can make it. Once the lambs are born, the ewes with one lamb are moved through a gate on to the field next to Morte Point which is their ultimate destination. The ewes with a pair of lambs head through a second gate, so that they will remain on the slightly better ground of Town Farm. The only treatment en route is to ring the tails of the lambs, treat their navels with iodine, and to put a small notch in the ear of the ewe-lambs to make a permanent record of which breed of ram was their sire, and thereby avoid in-breeding in years to come.

'Can you walk them down to the gate?' I called to Nick and Clare.

Soon, underneath a perfect pale blue spring sky, the four of us were waving crooks and running to and fro, walking the sheep in their various directions. The children seemed in their element. I certainly felt that way.

With Gail on maternity leave for a few weeks, I had brought Greg and Swift out this morning. As my most experienced dogs, they are quite capable of dealing with even the most overprotective mothers. I sent the pair of them beyond Laura, who was making good progress carrying a lamb while its mother followed, and they picked out a ewe at the far end of the field, who was sheltering her lamb away from the rest of the flock, under an overhanging hedge on top of the wall.

It's a real testament to the intelligence of Greg and Swift that they understand that, for this short period of the year, they are to pick out only the sheep with lambs. Completely ignoring the rest of the milling flock, first Greg, then Swift, homed in on the ewe and lamb.

The new mother was wildly protective of her new offspring. Turning to face the dogs, she lowered her head and made repeated jabbing attempts at whichever was the nearer. However, their experience was too much for her. An expert sidestep from Swift, then more forward movements, put the ewe into reverse mode, and soon, with the lamb at her side, she was being forced back across the field towards me.

'Got one for you, Daddy,' came a call from behind me. Nick was carrying a lamb, clutched awkwardly over the crook of his arm. He was trying desperately to hold it up and out of the way of the old ewe that paced around him. She knew full well where her lamb was, and was quite prepared to climb over Nick in order to retrieve it.

'Well done, Nick,' I said, relieving him of his load.

'I'll get the rings,' said Nick, running off to the Land Rover. Even at five, he was already earning his keep.

In little under an hour we'd successfully separated the flock into their relevant family units, and we made our way back across the field. 'I think you've earned yourselves something to eat,' I said as we shut the gates for the last time. Heading back to the Land Rover, my only slight worry was the two jet-black birds that Clare had disturbed as she had brought one of the last sheep down. The pair of crows were paying a little too much attention to the lambing field for my liking.

It is no surprise they gather, so rich are the pickings at this time of the year. Often they simply feed off the placenta, shed by the sheep soon after birth. This causes no problem at all, but unfortunately they have a nasty habit of taking their scavenging too far. In the first minutes after birth, new life is incredibly vulnerable, and crows soon learn that vulnerability can be turned into a choice meal, with newborn lamb's eyes and tongues at the top of the menu.

Although I hadn't yet had a problem this year, there had been odd instances in the past. On the farm of a friend on Exmoor, where large numbers of sheep are lambed on the hills, scavenging birds can cause losses on a large scale. It's the most miserable task on the morning round to pick up the remains of healthy lambs attacked in this way.

Nature is truly red in tooth and claw – and beak. For now I chose to do nothing, but if they turned into a real problem I would have to resort to the gun.

CHAPTER TWELVE

A Condemned Man

The voice on the end of the phone had a horribly officious and slightly superior tone to it.

'Mr Kennard, John Dennon from the Environment Agency.'

The last two words were enough to send me into an instant panic. The Environment Agency? What could they want? What had I done? It certainly couldn't be good news.

Mr Dennon didn't keep me on tenterhooks for long.

'We're undertaking a programme of on-the-spot checking of sheep-dipping facilities, to ensure that such facilities come up to the present European standards,' he continued. 'You do have a copy of those standards, Mr Kennard?'

I thought for a moment. I had a drawer full of booklets containing seemingly every rule and regulation ever applied to farming, but I sure as hell hadn't read them. I 'ummed and aahd' for a moment or two in an unconvincing manner, before deciding on a bet-hedging approach.

'Yes, of course, but it's some time since I've read them.' This was obviously the sort of foolish response that Mr Dennon was looking for.

'Good,' he said, a faintly victorious note in his voice. 'We have a team in the area this morning. They'll be along to inspect your sheep dip within the next hour, and I must warn you, the Agency is taking non-compliance very seriously.'

I felt utter panic.

The dip at Borough Farm is a little on the old side, to say the least. It performs the task of dipping sheep adequately and doesn't cause any problems, but there was no way in the world that it would conform to a list of rules and regulations long enough to fill a book. The government had recently given the Environment Agency awesome power. If I fell short of their exhaustive requirements, then I could probably expect a heavy fine, or imprisonment, or both.

'You can't come now,' I stammered, my crumbling mental state obvious. 'I'm in the middle of lambing, I haven't got time.'

To no avail.

'Under section 29 of the 2002 order on Farm Dipping Facilities, subsection D Part 2 . . .' the voice droned on. Here was a man who knew the law and certainly wasn't afraid to use it. At the end of his well-rehearsed lines he paused. 'So the team will be with you within the hour. Is that all right, Mr Kennard?' By now I was furious, but I had also begun to realise I was beaten.

'Well, I don't have any choice in the matter, do I?' I retorted without trying to hide my anger.

'No, Mr Kennard, you don't,' was the curt response.

I slammed down the phone without bothering to say goodbye. How was I going to get my sheep dip up to modern standards within the next hour? I didn't even know what might be deemed wrong with it. It could even be sited in the wrong place.

I made a dash for the drawer in the office where I kept my bulging supply of regulation booklets. Discarding heaps of unwanted folders on to the floor, I desperately looked for the relevant information. It wasn't there.

By now the panic was subsiding, to be replaced by a feeling of resignation. I couldn't do anything about any problems at this late stage; I might as well just accept the situation and in due course accept the punishment. Feeling overwhelmingly like a condemned man, I cast my eyes in the direction of the farm drive and waited for Mr Dennon's merry men to arrive.

I was in the kitchen, readying myself to go out to meet my execution squad when Debbie returned from outside.

'What's up with you? The children say you're in a panic about something.' She had lit the touch paper. In a moment I was off on a rant about the injustices of the situation.

'These people turn up in the middle of lambing, they just want to make a criminal out of normal farmers. How can I possibly conform with a list of rules and regulations as long as your arm, especially when I can't even find them!' I threw a handful of offending booklets on to the floor as if to make the point.

Debbie didn't seem to appreciate the gravity of the situation;

quite the opposite in fact. A hint of a smile was spreading across her face.

'You have remembered what day it is?' she said, one hand on a hip and a look of mild pity on her face.

'No, what do you mean?' I grumbled, my mind still busy imagining the grilling I was sure to get from the inspectors.

'The first of April.'

I just spluttered and stuttered for a moment.

'But . . . but . . . who . . . how . . . he said . . .'

Debbie quickly cut me short.

'Have you spoken to Andrew this morning?'

A glow of relief spread across my face.

'Andrew? But he couldn't have . . . yes he *could*,' I muttered to myself.

A couple of years previously, I had pulled off a rather successful April Fool's trick on Andrew. I'd sent him a letter, ostensibly from a local vegetable processing factory, offering him a job for his Shire horse on the carrot grading line. They reckoned Shires were the best judges of quality and wanted to draw on their expertise. Since then he had been threatening revenge, but never delivered – until now.

With the turmoil of lambing I'd completely forgotten what date it was, but even if I had remembered, I doubt that I would have challenged the frighteningly officious-sounding man on the other end of the phone.

When I rang Andrew's number a few seconds later, the phone was answered, but there was no voice at the other end, only the sound of fits of laughter. When the amusement eventually died down, I discovered that he had a friend staying, who had delighted in playing the part of Mr Dennon. I didn't know whether to laugh or cry, the relief of not facing the inspection was overwhelming, but to be so completely deceived was maddening. I hadn't been a condemned man, just a conned one.

'You should have heard your voice, when he told you that you didn't have any option,' said Andrew, who had been listening in on the conversation. The giggles were starting to return. 'Well,' he said, as eventually he regained composure, 'you have to have a laugh, don't you, David?'

'Yes,' I said through gritted teeth as I prepared to put the phone down, 'and you have to remember that there's always next year.'

CHAPTER THIRTEEN
Basic Instincts

There can be few finer places than the North Devon countryside in late April. Out on Morte Point this morning the landscape had been a mosaic of erupting colour: the pale green of the fresh shoots of bracken bursting through, the riotous hues of the sea campion, thrift and sheep's bit blossoming on the cliff's edge, and all framed against a sea so blue and passive it could have been the Mediterranean. It was a breathtaking place to visit, let alone work.

Back at the farm this afternoon the change of season was less dramatic but no less noticeable. In Borough Valley the ash and beech were dusted with pale new leaf, and the lush green grass seemed alive, the brightness of this year's yellow-orange dandelions rivalling even the brilliance of the flowering gorse.

The most obvious sign that springtime was well and truly here, however, was the swallows that had arrived home over the past couple of weeks. They now spent the early morning swooping low over the pasture, picking off the flies that had begun to swarm around the sheep. Their constant 'peep-peep' eclipsed the distant song of the thousand winged residents of the valley, that drifted up from the woods below.

As we gazed out across the rolling contours of Windcutter Down, I could see that Victor was more concerned with the distant dots of off-white grazing in the field on the horizon. 'Of course, if you had the right dog for the job, you could send it all the way to the top of that hill,' he said, a mischievous twinkle in his eye.

'Is that a challenge, Victor?'

'Might be,' he smiled.

He had rung the night before. He was passing by and wanted – as he put it – to 'see if you've got anything to compete with my Lad'. Victor, like me, had begun turning his mind to the return of sheepdog trialling in the coming weeks.

Although I only compete in the odd trial, I still felt excited at the prospect of getting involved once more. In the past Greg and Swift had been my main dogs, but prior to the outbreak Fern had begun to show real promise as a trials competitor. Her instincts were good and over the past few months there had been signs that she was losing some of the apprehensiveness that was her greatest weakness. It was she who stood next to Victor and me, eyeing the small flock in the field beyond Windcutter Down.

Even though they were more than half a mile away, the

rolling contours of the upper fields meant that they were clearly visible. To get to them, however, a dog would have to run through two gateways and round a third field. No easy feat. I felt confident Fern was up to the challenge, however. If she could work as well as she had been over the last few months, then I should be able to take some of the wind out of Victor's sails.

'Well, the older ones do that all the time, so why don't we give the pup a chance,' I said, employing a little kidology.

I called Fern to me and set her to my right, asking her to look as I did so, but immediately something about her demeanour told me that this was not going to be a good idea. Instead of focusing intently on the direction in which I needed her to run, she started to look around, as if confused. It was quite obvious that she had no idea of where she was meant to be going.

'Come bye, Fern,' I said gently. For a moment she pricked her ears and looked, then, as if to let me know that she hadn't the foggiest idea what I was on about, she tucked them down again and lay down at my feet.

Victor raised an eyebrow, but said nothing. I tried again, more firmly this time: 'Come bye, Fern.'

Nothing but a quick wag of the tail as she rolled over on to her back.

'What were you asking her to do exactly, David?' asked Victor with the first hint of amusement.

'Fern, come bye,' I said, a fraction of a decibel louder this time, but careful not to lose my cool.

This time Fern got the message. She sprang to her feet and, looking around wildly, she ran across the field in vaguely the right direction. Sadly, once she reached the opposite hedge she lay down and tucked her ears back, having completed less than a hundred yards.

From then on nothing I did or said could persuade her to her feet.

It has to be said that Victor controlled himself very well. He even showed a hint of sympathy as we talked briefly about the possible reasons for Fern's occasional aberrations. Deep down, however, I knew a small part of him was thinking contentedly: 'No competition from that department this season.'

I tried to change the subject.

'We've been thinking about some sheepdog displays up here,' I said, eager to see his reaction.

'Hmmm,' he said, looking genuinely taken aback. 'Where did you get that idea from?'

Since Derek and Helen had first planted the thought in my head during my visit to Cumbria, it had refused to go away. Debbie and I had spent a long time talking about it over the dinner table at night and our discussions had done nothing to dampen our enthusiasm. We felt sure that Woolacombe had more than enough visitors over the midsummer months. What made it even more appealing was that the main route into the village passes close to the end of the farm drive, which would, we hoped, make it more likely that visitors would find us.

The fields at Windcutter Down seemed to offer the perfect setting as well. It was this latter point that Victor saw as a strength.

'Well, you couldn't beat it for scenery,' he said as we stood and scanned the view for a few moments. 'How many people do you reckon would turn up?'

'I've no idea, but if we don't give it a go, we'll never know, and it's got to be worth trying with the sheep business the way it is. Have you sold any lambs yet?'

Victor's farm is on the warmer, more sheltered side of

Devon, consequently he can produce lambs for sale earlier in the year than I do.

'No,' he said, with a gentle shake of his head. 'I've got them to sell, but the price is still low. There's just not the demand in the market. Earning a few quid away from farming has got to be the answer.'

We walked on back down the drive. As he got ready to leave, Victor climbed into his Land Rover and wound down the window.

'When are you going to start these displays then?' he asked.

'Haven't really set a date yet, but perhaps on the Bank Holiday at the end of May. Do one then and see how it goes,' I said.

He raised an eyebrow, as he tried to conceal one final mischievous smile.

'You'll be wanting to borrow my dogs for the day, will you?'

————

Clare arrived in the yard, with a face like thunder.

'I can't believe those stupid ducks,' she said. 'I've been waiting all year for them to start laying, and now they have, they've laid their eggs right at the bottom of the pond.'

She hauled me off to show me the cause of her frustration.

Now that they were twelve months old, Clare quite rightly had expected them to start laying, but they had resolutely refused to do so – until today. Clearly visible dotted around the bottom of the pond were half a dozen duck eggs, shining white, but all completely inaccessible to anyone who didn't fancy a short dip.

'Dad, what are we going to do about it? We can't waste all their eggs.'

I was working out how I might build a platform that would

be accessible only from the water, that might give the ducks a suitable place to lay, when we were interrupted by Laura.

'Dad, I think Gail's going to have her pups; she's gone behind the washing machine, chewing up pieces of paper.'

As her pregnancy had entered its final weeks, Gail had taken up residence in the kitchen. It wasn't that she didn't want to go to work – she still tried to escape to join me outside – but her condition simply didn't allow it. Her physique was for the time being simply not conducive to the job of a working sheepdog. She panted her way around the house looking about as uncomfortable as a pregnant dog could look, invoking much sympathy from Debbie and our female visitors.

Laura was right. Gail's nesting instincts had taken over, and she had disappeared into the most inaccessible crevice in the entire house, but with a little effort, she was soon coaxed back to a more appropriate bed at the foot of the dungeon stairs.

That evening, under the constant and watchful gaze of Debbie, the children and me, she produced nine pups: five bitches and four dogs. It was an amazingly large first litter, but Gail was equal to the task.

The way nature invokes the maternal instincts of a new mother is nothing short of a miracle. All evening Gail worked tirelessly, washing and grooming her new family, nudging with her nose to ensure that each pup was in a position to find its first life-nourishing feed. As she did so, slowly the high-pitched whimpers of the ravenous pups subsided. Then, finally, Gail laid her head over the edge of her bed and shut her eyes.

It was a picture of utter contentment.

————

When I heard the drone of motorbike engines drifting up the valley early one Sunday morning I felt the red mist descending immediately.

On two of the previous three Sundays, scramblers had torn around the woodland at the upper end of the valley, scarring the landscape as they did so. I hadn't seen or heard them, but the muddied tracks that they left behind made their presence all too obvious. I find it difficult not to get this sort of vandalism out of proportion. Here was some of the most beautiful woodland imaginable, an almost virgin landscape where populations of bird and mammal life thrived. Seeing the royal-blue haze of bluebells ripped apart by the tracks of motorbikes was sickening. It was like seeing graffiti scrawled on Constable's *Haywain*.

So today I was fired by adrenaline as I set off to catch the culprits. It wasn't a difficult task. I soon found two scrambling bikes ploughing merry channels through bluebell-clad banks. One rider was just attempting a particularly tricky manoeuvre as I approached, trying to thread his bike between two whips of sycamore on the steep slope. As I watched, he stopped, overbalanced and slid gracefully down the bank, arriving neatly upside down at my feet.

If you are a leather-clad, helmeted scramble bike rider, you probably imagine your appearance is menacing enough to ward off all potential threats. However, any gladiatorial advantage that the lad might have had was negated by the fact that he met me first from the feet upwards.

Slowly he hauled himself upright. In fact, as the second rider came over to join us, it turned out that they were no more than seventeen. They confessed to being local and, in fairness to them, muttered some form of an apology when I made my feelings clear about the effect they were having on the landscape of the valley.

Eventually they rode off up the track and out of the woods, promising not to return. I had a little sympathy with them. I could well understand the exhilaration of scrambling; it's just that it has to be in the right place. The sad truth was that the pair of them were only reflecting the view of what seemed to be a growing majority of people: that the countryside is merely a place for urban recreation. Walkers have always been a part of the landscape, of course, but intrusive forms of recreation were becoming more and more common. In recent years, the peace and solitude of our family's Sunday excursions on Exmoor had been ruined by columns of scrambler bikes and four-wheel drive trucks churning the mud. For a few moments of excitement the landscape is left scarred, sometimes for years, and pockets of true rural tranquillity are becoming increasingly rare.

Debbie and I had decided to put on our first sheepdog display at the end of the month, over the May Bank Holiday weekend. It made sense to give the idea a trial run and this weekend was usually a busy one, drawing healthy numbers of holiday-makers to North Devon.

'What have we got to lose?' Debbie had said, showing an admirable positive spirit. 'If it comes off, great; if it doesn't, we'll have an idea what we could do better in the summer.'

I hadn't been able to fault her argument and had begun preparing.

With the display now looming into view my training sessions with the dogs had taken on a new level of importance. Having dependable dogs has always been important to me, but, if I'm honest, I've always felt a little guilty about spending extra hours on the 'polishing' that's required to compete at sheepdog trials. I'd always restricted sessions to the long, light evenings of the spring and summer. Now though, I had every justification for my evenings in the fields. If we were to impress any audience that we managed to attract, then the dogs would have to be on best form, Greg and Swift in particular.

I had another reason to start stepping up the practice work. For the second time I had entered them to run in the brace competition in the English National Sheepdog trials, the premier event of the trialling season in England.

I couldn't help but feel that, as the years had gone by, their abilities as a pair had waned a little. Swift was still at her peak, and I had no doubt that she would be capable of performing at that level. It was Greg who concerned me. He was beginning to show his age, with some justification. He did, after all, have many years of hard work behind him. Yet, if I didn't ask too much of him over the next few months, then I was sure he would be up to the task. A little brushing up over the next six weeks was all that was required. As I set the pair of them off up Windcutter Down a few minutes later, however, I wondered whether I needed to bother. At their best, Greg and Swift really were a dream team.

With no whistle or word of command from me, the pair of them took up position at the far side of the sheep in the uppermost corner of the field. Greg was on the left, Swift on the right. This was the way that they always worked, a long-formed habit.

Without fuss, they brought the few sheep back to me. Then I blew a whistle to stop Swift while Greg worked. After a while I reversed things, stopping Greg while Swift took over.

Asking a dog not to work the sheep in front of them is difficult, but this evening shepherd and dogs were in harmony. I felt pride and satisfaction, but it was quickly replaced with a nervousness. I bet they don't go this well in the first public display, I thought.

After running through a few more routines, I called the pair of them to me. Greg's tail broke into a great wag. He knew that I was pleased with him, and bounded over, leaping about me. Swift made a more sedate approach and nuzzled my hand.

'Just make sure you behave like that in a couple of weeks,' I told them as we headed back to the Land Rover.

I was about to climb in when the lolloping oversized pup in the back suddenly sat up and looked at me. Up until this point Jake had been chewing on a small branch that he had managed to hold in his mouth as he jumped in, back in the yard. Now he suddenly looked interested and alert.

I'd been a little reluctant to try Jake on sheep. His playful, eccentric behaviour didn't fill me with confidence. I didn't think he was going to show any interest in working. Indeed, much as I enjoyed seeing him accompanying Clare around the farm, I was beginning to suspect he might prefer life as a companion dog. Looking at his animated features now, though, I thought this was as good a time as any to find out for sure.

'About time you had a run with some sheep,' I told him as I lifted him under my arm and carried him towards the flock still standing nearby.

I've introduced many a pup to sheep for the first time and knew it was a moment of wild unpredictability. Some pups head straight into the middle of the flock, sending sheep scattering amidst a snowstorm of wool. Others only show interest for a few seconds, before they wander off sniffing at more enticing smells.

Jake, however, showed me something entirely different. As soon as I put him down he focused completely on the sheep ahead. As I made encouraging 'whooshing' noises he headed for his quarry with purpose. With the look of a pup well educated in the art of herding, he approached the sheep, and proceeded to run around them, with a staggering level of self-assurance.

'Good lad, Jake, good lad,' I whispered as he held the sheep together, never once making a move that might cause one sheep to break from the rest.

This was the sort of work from a first-time puppy that I dreamed about. It was the equivalent of having a child with a natural gift for music. It all seemed to come so easily to him. When, a few minutes later, I called him back to me, he just looked up and trotted over, with a great wag of the tail.

That look he gave me was one that I had seen only a few moments before. I spoke to him softly, while stroking the top of his head. Perhaps this really was Greg the second in the making.

CHAPTER FOURTEEN
'The Best Days of Our Lives'

Driving along the lanes towards the village of West Down, the trailer containing my shearing equipment rattling noisily along behind the Land Rover, I couldn't help feeling an era was drawing to a close. I'd worked with the same shearing gang since I'd arrived in Devon. With my colleagues Geoff and Chris I'd spent each May and June working a round that had been passed between shearers for more than thirty years. It hadn't made any of us wealthy, but it had provided an extra income of sorts.

It had always been something of a miracle that we squeezed in the time to shear in the region of fifteen thousand sheep each season. Geoff fitted shearing in between milking two or three herds of cows, tree-surgery and a variety of other farm-related jobs. I too had found it increasingly difficult to spend long days away from the farm. However, it was Chris who had finally found the toll too much. With his father, Gerald, he farmed a mixture of sheep, cattle and corn near West Down, working from dawn till dusk to keep it a going concern. It had always been a source of amazement to me that he not only found time to shear but was also the gang's organiser. Now, with Gerald approaching retirement, something had to give and Chris had reluctantly announced his intention to leave the gang at the end of the year.

As it turned out, the first week of shearing had been uncharacteristically easy: fine weather had seen half a dozen of the early flocks shorn. So today we'd taken the opportunity to clip off Chris's own sheep.

Help is always at a premium on shearing days, so as I pulled into the yard I was pleased to see a healthy band of assistants gathered outside Chris's sheds. Gerald was there, as usual, but it was only when I looked a second time that I realised the two towering figures at the back were Chris's sons, Tim and David.

'Crikey, Chris, what've you been feeding them on? They've doubled in size since last year,' I said, hauling my kit off the back of the trailer.

'It's the silage they eat!' said Chris dryly.

In years gone by I'd ribbed the two boys mercilessly, but now, looking at the size of them, it no longer seemed such a good idea. I'd have to turn my attention elsewhere today.

It wasn't long before a natural replacement was looming into view. As I set up my kit alongside Chris and Geoff, yet another pair of helping hands pulled up in the yard.

'Jus' come to make sure you boys is gettin' on with the job,' grinned John, a neighbour of Chris, who often helped out at this time of the year. I'd never yet seen him without a cloth cap sitting jauntily high on his head, and a roll-up cigarette drooping from one corner of his mouth.

John was one of the old breed of farmers, whose solution to every downward turn in the industry was to work harder. He'd started work at the age of twelve. Now in his late sixties, he and his son had bounced back from losing their stock during foot and mouth, and were now farming a larger flock than before to make up for the dwindling prices.

As a lifelong sheep farmer, John believed he knew a thing or two about sheepdogs, and certainly far more than some 'foreign' upstart like myself. So, on the rare occasions we met, he'd often give me the benefit of his knowledge, in his own irrepressible way.

'Got a dog in the van 'ere better than anything you ever 'ad.' John always wore the same straight-faced expression, making it almost impossible to tell when he was being serious or when he was having a joke at your expense. On top of that he spoke in a Devon accent so heavy it would need translating if he ever wandered beyond the county's boundaries.

'Where did you get it from, John?' I asked.

'Bought it from a bloke up over,' he said, using the local expression for Exmoor. 'Does anything, 'e does, you'll not of sin anything like it.'

'Get him out the van then, John; he can help get the sheep in the shed,' I replied with a knowing smile.

With a little more prompting, John retreated to the van, and returned with a sheepdog at foot. It was soon apparent that whatever schooling the animal had received in its youth, it was going to require a little more 'readjustment training' in order to adapt to its new owner. For the next few minutes, there was a flurry of commands – "ere dog, 'ere dog . . . lay down, lay down . . . bloody dog' – all delivered with John's roll-up still firmly fixed in the corner of his mouth.

The dog scurried back and forth across the yard, splitting sheep in all directions. Eventually he came to a stop, lying down some way from his new master then looking up with an expression of thorough confusion.

John, however, was undeterred.

'Bes' dog you'll ever see,' he repeated without a hint of embarrassment. 'Won't work for me, mind, only for the bloke what trained 'im.'

———

Soon the clippers were running and wool was flying, the extra hands making for lighter work. Tim filled the pens and caught as many sheep as he could for the shearers in between times. David wormed the lambs, while John and Gerald rolled the wool.

The warm weather had made the sheep sweaty and perfect to shear. These were the sort of conditions that could make any shearer look polished.

Unlike Geoff and Chris, I never claim to be an expert in the art of shearing, merely proficient. On a good day I could clip a little over two hundred sheep, a respectable amount but nothing compared to the real masters of the art who can clip five or six hundred in a ten-hour day, an average of one every forty seconds. The world's greatest shearer is a New Zealander called David Fagan, who has won the world shearing title, the 'Golden Shears', on eleven occasions. Given the skill, stamina, strength and fitness required to shear to that standard, it's a sporting record that in my opinion stands alongside that of Sir Steve Redgrave.

There were no records to be broken today, however, and when coffee arrived, courtesy of Chris's wife Doreen, I gratefully slumped back on to one of the now fully stuffed woolsacks.

The first days of the shearing season are always the hardest physically. Your body has simply forgotten the pain it has to go through, but by the time you've spent the first twelve hours bent double over a procession of truculent sheep the memories come flooding back. I could feel the pain welling up in every muscle already.

It was clear Geoff was feeling it as well. He stood up, arching his back painfully.

'You look like a pregnant duck when you do that, Geoffrey,' I told him, completely unsympathetic to his troubles. He responded with a torrent of abuse, largely about the lack of speed I'd shown this morning. Geoff had shorn three more sheep than me.

As the early afternoon sun moved round to the south of the barn and shone directly on to the assembled gang, the sweat flowed more freely, but with the extra heat the stream of gleaming-white sheep leaving the barn became a torrent. At a quarter to nine that evening, Chris, Geoff and I each wearily pulled our last animal of the day from the pen. One by one the motors fell silent, and the last sheep ran from the shed door.

'Is it ever worth it?' Chris muttered quietly a few moments later, looking every bit as shattered as I felt.

The current rate for shearing was 70p a sheep, precisely the same as it was when I first picked up a pair of clippers twenty years ago. It was little wonder so few youngsters were interested in learning the art. There were a million and one easier ways to earn money.

Geoff didn't answer. His eyes were screwed up and his back arched once more. He was really suffering.

The freshly shorn sheep were by now heading out of the yard, and a peace returned to the shearing shed. There was something satisfying to me about the whole day.

'I know the money's not much any more, but we're fit and healthy, and the company's good,' I said. 'One day we'll look back on these as the best days of our lives.'

The look they both shot me spoke volumes. They clearly thought I'd gone mad.

———

'Perhaps if these displays take off, you'll be able to give up the shearing,' Debbie said with what sounded like a hint of impatience. With me lying prostrate on the kitchen floor, she was trampling over my lower back. This was the only treatment that I found gave comfort to the pain caused in this particular area.

'I can't really stop. Chris is packing it up this year, and Geoff will have no one to shear with,' I said, slowly picking myself up off the floor a moment or two later. 'Besides, I'd miss the company if I stopped, and it does you good to get off the farm for a few weeks a year.'

I leaned against the Aga then finally straightened myself upright. 'But I must admit working a sheepdog doesn't make your back feel like this.'

———

The display we'd planned for the Monday of the May Bank Holiday weekend had become a dominant topic of conversation between Debbie and me.

We'd spent an age fretting over the wording and design of the posters, opting in the end for something straightforward. It had been only in the past few days that we'd distributed them, and Debbie had got a few sceptical looks when she'd asked shopkeepers and campsite owners to stick the notice up somewhere prominent.

I had been torn as to what sort of response I wanted for the first display. In a way the embarrassment of no one turning up wasn't as bad as the idea of a coachload of tourists pulling into the car-park area I was going to rope off on one side of Windcutter Down. The very thought of having to perform to dozens of people was enough to tie my stomach into knots – and if it did go badly, the fewer people that knew about it the better. So in the end Debbie and I had decided to put a board up at the end of the drive on the day, and stick up just the half dozen posters in the village.

'It's supposed to be one man and his dog, isn't it,' I'd joked rather limply.

CHAPTER FIFTEEN

One Can't Start, One Can't Stop

J ust after ten in the morning, and the Dartmoor sun was already high in the sky, adding a burnished, golden glow to a landscape that stretched twenty miles and more towards the great Tors in the distance. The breeze was soft and suffused with the sweetness of flowering hedgerow and spring grass. A more perfect setting for the first sheepdog trial to be held in the south-west since foot and mouth would have been hard to imagine.

After a quick check of the stock, I'd set off that morning quietly excited at the prospect of the day ahead. In the back of the Land Rover, Fern was accompanied by Ernie, who, despite knowing nothing of what lay in store, seemed to share my mood of anticipation.

I've never been entirely sure what it is that makes trialling so strangely compulsive. Partly it's an enjoyment of the company of my fellow sheepdog enthusiasts, many of whom are shepherds and share a unique understanding of the curious life we lead; partly too it's a love of the sheepdog and its working relationship with man – but, to me at least, it's also a sport rather than a hobby, and for that reason brings all the satisfaction, and frustration, that winning, losing and sometimes just competing bring with them.

It may not seem it to the untrained eye, but sheepdog trialling is as competitive as any other sport. Triallists are there to win, not to make up the numbers. Today, however, there was a real sense of gratitude that at last the sport had returned.

Parking the Land Rover, I bumped into Tom, one of the pool of regular competitors in the southwest. Since taking retirement a few years earlier, he had managed to get to just about every trial each summer.

'What did you find to do with yourself last year with no trials to go to?' I asked with a smile.

Before he could answer his wife Joan was leaning out of the window of their car. 'All those jobs that he hasn't been doing for years, because he's been too busy trialling,' she smiled.

As if to prove the significance of the competition, a strong contingent from Wales had made its way across the Severn Bridge for the day. While the Southwest of England trialling just about holds its own, with between twenty and thirty regular competitors, Wales has many times that amount. In the rural areas across the country, the sport is one of the most avidly followed events in the diary.

I was once chatting to a Welshman at a trial. 'I usually get to about a dozen trials in a year,' I'd said to him by way of conversation. He looked at me a little surprised.

'Crikey,' he said, 'I do more than that in one week in August!'

Several of the faces I now spotted were regulars in the Welsh national team. At the highest level, sheepdog trialling is no different from any other sport: it takes a huge amount of skill, knowledge and dedication to do well consistently. However, all the ability in the world is of little use if you don't have the right dog. The skill of the best handlers is that they also seem to be able to breed the right animals.

The line up of dogs tied up to the fence behind the parked cars today represented some of the best dogs in the country. They were all Border Collies, but covered a cross section of colours, markings, sizes and appearance. None of them were quite as 'unique' as Jake, I thought to myself with a smile.

———

'Can't think why you've turned up,' came a greeting from behind me.

I turned around to see Victor, stoutly holding his crook, with his sharply marked, short-coated bitch Tammy sitting compliantly at his feet.

'You next to run, Tammy?' I addressed his dog. 'I'll work you today if you like. You're bound to get on a lot better.' Tammy totally ignored me. She always did when I made my numerous attempts to lure her to a better life in North Devon. She was a tremendous sheepdog.

The previous run came to an end and Victor and Tammy

strode purposefully forward. A word to the judge to confirm names and they were off. I settled back with a few other familiar faces to watch.

Tammy sped off to her left on the outrun and away towards the sheep at the bottom end of the field. Most sheepdog trials follow a similar pattern. The competitor stands at the post and sends his dog to collect five distant sheep. The length of this outrun varies considerably, and it is generally considered that the further the sheep are away the more difficult the course is to master, although there are other factors that add to the difficulty. The whole trial should reflect the sort of everyday work required of a sheepdog, and courses are often on rugged country, with rough ground, walls and banks to negotiate. Dogs are sent off with no more than their owner's commands to guide them, a real test of partnership between shepherd and dog when there may be half a mile of outrun to cover.

Tammy arrived at her sheep without a problem, and started 'fetching' them back up the field. Here the sheep have to be brought in a straight line, passing through a pair of hurdles seven metres apart, all the way back to the handler. Then, once the sheep arrived back at Victor's feet, the 'driving' section of the trial began.

A couple of whistles, and the sheep turned around, and started driving away from the shepherd, towards a pair of gates a hundred yards to the left. Up until now, things had been going well, to the point where the chatting huddles of onlookers turned to give the events on the course their full attention. The sheep, however, had detected a chance to escape, and were gaining speed rapidly as they approached the first drive gates.

This is always a dilemma for the handler. If you leave the dog trailing behind, then you have little hope of guiding the sheep through the gates, but if you tell the dog to move on quickly behind them, the sheep will go even faster.

A whistle from Victor, and Tammy was gaining on the galloping sheep. A quick left and a quick right, and the five swerved a little before heading straight between the two hurdles. However, the next task was for Tammy to turn the sheep into the 'cross drive' and through another set of gates. Now the sheep were running at such a pace that they had gone fully fifty yards before she could turn them back in the right direction.

What had promised to be a great run was suddenly looking mediocre as Victor did battle with five sheep that had become suddenly rather less compliant. After shaving the wrong side of the second set of gates, it was back to the shedding ring, where two sheep have to be separated from the other three. Tammy was faultless here and completed the run by putting the five sheep into the small pen with some style.

Sheepdog triallists are normally conservative in their appreciation of a completed run. It takes something quite exceptional to break the air of indifference that the penning of the sheep usually creates, so a few seconds later Victor duly walked back to the usual deafening silence that the conclusion of a run generates. I knew he would be giving me a ribbing soon enough, so I had no qualms about getting in first.

'Why were you trying not to get them through those gates over there?' I asked. We analysed the problems he'd had with the sheep, then Victor turned his attention to me.

'So which of your dogs are you running today? It doesn't say in the programme.'

He looked slightly shocked when I told him.

'It's about time I tried Ernie out; and if Fern is going to

make it, she's going to have to start doing well this year,' I said, trying to appear confident.

Victor shot me a quizzical look. He'd seen Fern on the farm and knew her problem. He'd also heard me talk about Ernie and his idiosyncrasies too.

'So you're running one dog that you can't start, and one that you can't stop? This'll be interesting,' he said, after a pause.

I laughed, 'Interesting might be an understatement, Victor.'

———

The general sense of frivolity continued throughout the day. I spent a pleasant hour chatting with friends, until someone mentioned that it would be my turn to run in a minute. Ernie was the first to go.

Running Ernie had seemed a good idea twenty-four hours ago. It was something I'd been thinking about for several months, and at home last night I'd convinced myself that the time was right, but as the two of us made our way to the post, I couldn't for the life of me remember how or why I'd come to that conclusion.

Positive thinking is the key to all sporting success, so it didn't bode well that my head was filled with bad memories of Ernie's headstrong behaviour. There was the day he'd rounded up sheep in the wrong field, and the morning he'd sent me into the ice-cold waters of Morte Point. Thinking about it, there were so many times when he'd ignored me and followed his own instincts. By the time I sent Ernie scampering down the hill, I'd all but talked myself out of any thoughts of winning. Keeping Ernie under my command for the next few minutes would be triumph enough, I told myself.

As it turned out, he started reasonably well, picking up

his sheep without a problem. Encouraged by a combination of forceful whistling and growling from me, he then brought the sheep towards the fetch gates well enough for a degree of optimism, but still I could sense that inside he was a coiled spring. As one sheep pulled too far to the right, he was off in an explosion of pace, and in turning it back, upset the other four, causing all five to miss the gates.

'Ernie, stop . . . Stop . . . Stop,' I growled, and eventually he did.

Still I kept on, using my voice to add weight to the whistle. As both he and the sheep grew closer, he became more responsive to my tense voice, working more and more slowly – until on the first section of the driving, he obviously decided that, with the boss in this sort of a mood, he would have to go on a 'go slow', and proceeded from then on at half speed, moving only when told, and even then at a snail's pace. The sheep responded to the lack of canine pressure by grazing as they walked. Although we successfully negotiated the next two sets of gates, with reasonably straight lines en route, it was no surprise to me when, long before the sheep entered the shedding ring, the bell sounded for the end of my allotted time.

As we drove the sheep from the course, I could guess what had been going on in Ernie's mind. He had been desperate to kick into top gear, and get the sheep scurrying along as he thought they should. I, on the other hand, wanted him to show more control, accuracy and patience. Frustrated by all my grumbling and whistling, he'd taken on the persona of a sulking child. Before I trialled him again I would have to find a happy compromise.

An hour or so later, the dog that I still hoped everyone would come to fear was standing next to me by the post. Having trained and worked Fern over three and a half years, I could now read her expressions like a book, and for once she appeared to be focused and intent on running in the right direction. I started to feel a buzz of confident anticipation. I knew that once she arrived behind the sheep we were in with a chance.

With a 'Come bye, Fern' from me she was off at full speed, which was a good sign. She was running in the right direction as well.

Perhaps this is the breakthrough, I thought to myself as the sharp black figure disappeared from view over a small rise in the ground. I regretted it almost immediately.

It's always nerve-racking when a dog is lost from sight momentarily. There are small signs that one looks for on these occasions, perhaps a bird taking flight to indicate the dog is passing. I could still see the sheep in the distance, so I was hoping to see them turning towards Fern as she approached, something that should have begun to happen within thirty seconds or so of her disappearing. However, thirty seconds soon turned into a minute, producing the first pangs of panic. I gave a loud right-hand whistle, but there was no sign. The confidence of earlier was already ebbing away. One minute turned into two. I gave another loud blast, but still nothing. I could feel the weight of the gazes on my back. A few more calls and blasts of the whistle and I'd have to walk to look for her. Humiliation.

Just as I was about to make that awful decision, she appeared, on the other side of the hill away to my right. She had crossed a fence and was now eyeing the sheep that had been dispatched there after the previous runs.

'Fern! That'll do . . . that'll do.'

Reluctantly she sloped back towards me, finding a spot to squeeze under the fence en route. Once she had reached a point a hundred yards away, I sent her again, in the hope that she might find the correct sheep at the second time of asking.

'Come bye, Fern,' I said, with a little more feeling.

This time she ran properly and arrived behind the sheep as she should have done in the first place, but the run was already ruined. The judges would have taken all of my outrun points, and I was lucky that they didn't ask me to retire. I pottered Fern and the sheep around the course, feeling that the experience would do her good, but inside I was mortified.

This was the same problem I'd been having with Fern since I'd first started training her. I thought we'd made good progress in eradicating it, but clearly we'd taken one step forward only to take two more backwards. I gave her the usual rub behind her ear and wandered back towards the truck.

Luckily, the frustration I felt didn't last long. As I headed back northwards across Dartmoor and some of the most spectacular views in the country, I remembered what today had been all about, and I began to feel content simply at having been part of it. Today hadn't been about individual dogs but about trialling, and this ages-old aspect of country life returning to normality at last. That was a cause well worth celebrating.

———

Just after three, I drove to the end of the drive, and – desperately hoping that no one I knew would catch me in the act – hammered a white sign into the grass verge. It read: 'Working Sheepdog Demonstration Tonight, 6 p.m.'

Actually, it read 'Working Sheepdog Demon stration'. Three-quarters of the way through painting the sign I'd realised I hadn't left enough space for all the wording. So, rather than starting all over again, I'd attached an extra piece of board, which had created a gap. Let's hope there are no sheepdog demons here tonight, I thought to myself as I stood back to inspect my handiwork.

I didn't hang around for long. I knew I'd get a ribbing from friends when they found out about my attempt at diversification. The board would only have to be up for a few hours, and I hoped no one I knew would notice it in that time.

The sign might have been less than perfect, but the same couldn't be said of the site on Windcutter Down. It looked an absolute picture.

That morning I'd laid out a few straw bales in a semi-circle around the area where most of the action would be. There was a pen a few yards in front, where I hoped that the sheep would end up. At various places around the field I had placed wooden gates, a few yards apart, through which I intended to drive the sheep. Set against the backdrop of the Borough Valley, it all looked quite impressive, particularly given the wonderful weather. Apart from the odd wisp of cloud at high altitude, the sky was a flawless blue.

With everything in place, I decided that the dogs had better have one last run. Greg, Swift, Gail, Fern and Ernie were in the back of the Land Rover and I let them all loose at once. It's often said that animals can pick up on human nerves, and it's certainly true of sheepdogs. I'd seen it on more than one occasion when trialling. I don't know whether they sense the trepidation in general or there's something about the way I issue my commands when I'm nervous, but today they'd definitely spotted my anxiety. So I tried to talk to them in the most gentle and reassuring tones possible.

By 5 p.m. I'd finished scribbling a few notes on to a piece of card. Public speaking wasn't exactly a regular part of my day-to-day life as a shepherd, so I didn't really have much idea what sort of patter I needed, but the few notes were my best attempt and would have to do. The final job now was to disperse a dozen sheep into release pens that I had erected at a couple of locations around the field.

I used Ernie to pen the first bunch, and they went in without too much trouble. The second group posed more of a problem, however. They turned and twisted, trying to break past him. One ewe in particular had a look in her eyes that told me she had no intention of playing ball.

Ernie's turn of speed was superior to that of any sheep, but controlling them is about much more than that. His instincts can be too strong. Too often, when he should have been staying still, he was moving around. Time and again, just as the sheep were about to turn into the pen, he would make a move that would hinder rather than help.

If you couldn't put the sheep in a pen in the corner of the field with no one watching, how are you going to do it with an audience? This is going to be a humiliation, I was thinking.

Eventually, the wild-eyed ewe conceded, and turned in. Ernie panted and looked up at me. He seemed unduly puffed for such a short piece of work.

'We'll keep you in reserve for this evening, I think, Ernest.' I didn't want to take the risk of him losing the plot in front of the audience.

By the time Debbie and the children arrived in the field it was 5.20. Forty minutes to go.

Nick was going through one of those phases where his conversation consisted almost entirely of questions.

'When are the people arriving?' he asked.

'Good question, Nick,' I replied, 'and the answer is, I don't know.'

The fine weather had brought bumper numbers of visitors

to Woolacombe. The mini heatwave we'd been enjoying had even sent the hardier ones into the sea. Now there was a constant stream of cars heading along the road at the top of the drive, away from Woolacombe. None of them seemed to notice the sign, let alone slow down to turn left into the lane.

By a quarter to six I was beginning to feel despondent. The children seemed a little confused, unable to understand why we'd gone to all this bother, if no one was going to come. Debbie was still trying to be positive, but her face told me that she, too, was now fearing the worst. I kicked the bottom bar of the gate in frustration, and began in my own mind to reappraise the whole situation. I wouldn't be able give up the shearing at all. It had been a ridiculously over-optimistic idea, to think that there was money to be made from a display with sheepdogs.

I was about to tell Debbie what a waste of time the whole thing was when a car appeared in the drive with a large family and their small dog crammed inside. Our first customers. Debbie gave a smile as the driver wound down his window.

'Is this the right place for the sheepdog display?' asked the father, clearly thrown by the fact there was a field, dogs and some sheep – but no audience.

'It certainly is, you're the first to arrive,' said Debbie. 'You can park in the corner of the field.'

We'd decided that we'd only charge a nominal fee for the display. These customers didn't seem to have any problem with it as the father happily handed over a few pounds. Hardly a runaway success, I thought, still unable to shake off the feeling that this was a bad idea. I could have sheared seven sheep in fifteen minutes without all this hassle and still be better off.

A few moments later, however, a second car rolled up. It was

Gary and Glen, a retired couple from the village, collie owners and friends of the family.

'We thought we'd come along and see what you were up to!' Gary said, handing over their entry fee to Debbie. 'I have a feeling that one or two more might come along from the village as well.'

He was right. Over the next ten minutes half a dozen more cars rolled in, mainly locals, lending their support to our new venture. There was also a photographer who said he was looking for an interesting subject: 'I've only come along to take a few snaps,' he confessed.

By six o'clock eighteen people were perched expectantly on the straw bales.

Better make a good impression to start with, I thought, so I called Swift from the Land Rover. I was fairly certain that she would cause no problems. Of all the dogs, she was the one most capable of focusing on the work in hand to the exclusion of everything else going on around her.

'Well . . . er . . . good evening, everyone. Um . . . Welcome to Borough Farm. Thank you all for coming,' I started, hoping the nervousness wasn't showing in my voice. 'I'm going to demonstrate some working sheepdogs.' Silly thing to say, what else would you be doing? I thought. 'This is Swift; she's one of my best working sheepdogs, and if I blow a whistle like this she'll run to her left.'

I went to put the whistle to my mouth, but as I did so remembered that I hadn't released the sheep. Another mistake.

Fortunately Debbie had spotted this and scurried over to the pen to open the gate. I blew a left whistle for Swift and she was off. Immediately I felt relief that the audience's attention would now switch to the dog rather than me. Swift was living up to her name this evening and made a lightning circle around the sheep. At the sound of a second whistle she dropped to the floor, and on a third call she was off in the opposite direction. I looked across to the row of faces, and there was no doubt that they were engrossed, some even looked fascinated. The man with the camera clicked a few times and moved a little to get a different angle.

Relaxing now, I talked about Swift and her style of work. It seemed to be going better. I could see Clare and Laura hovering a little way away, unsure whether to look interested or embarrassed. Nick had no such inhibitions; he'd taken up a position on the end bale and was grinning broadly. At least I had one satisfied customer.

As Swift finished her routine, a heartening round of applause rang out. I opened the back door of the Land Rover and put her away, releasing Fern for her moment in the spotlight.

As a precaution, I talked a little bit about Fern and her good and bad points, before asking her to perform a shed and a few other manoeuvres, taking care to avoid anything that I thought she might object to. Another polite round of applause rippled across the fields as she came through without any mishaps. Slowly my butterflies were subsiding.

By the time I had put Fern away, Debbie had made her way to the far end of the field, where she freed the second set of ewes.

For this next part of the evening's entertainment I wanted Greg and Swift to demonstrate their brace work. With the two of them on good form, and on a field that they knew so well, I felt reasonably confident this wouldn't be a problem.

They hurtled off, as ever one to the left and one to the right, and a couple of minutes later, little more than a couple of black dots on the horizon, they arrived behind their sheep. The sheep started to move, but as they did so, the stroppy old ewe that Ernie had battled with earlier decided to remind me of her presence. Instead of ambling along with the rest, she started breaking, first one side and then the other, testing the resolve of the dogs.

As she came nearer, from the look in her eye it was obvious to all watching that this was a sheep that didn't want to play ball. I began to panic inside. My back was towards the visitors, and I couldn't gauge their reaction, but to me this ewe seemed intent on ruining my pièce de résistance. Valiantly Greg and Swift battled, but, just as I was attempting to complete the run by putting the sheep into the pen, she broke off, running at breakneck speed across the field whence she had come. This was the sheepdog demon, I thought to myself.

'Look back, Greg,' I called in desperation. In a flash he was off after her, but such was the ewe's pace that she had gone fully two hundred yards before Greg managed to pass her.

What followed typified Greg's mastery of his art. It was too far away for me or the audience to see it, but he must have fixed the ewe with a look that said: 'Give up, you're beaten.' The ewe accepted it immediately and within moments had turned and, like a sulky child, trotted back to join the rest. A few seconds later I shut the pen door, to the sound of truly hearty applause.

After fifty minutes I drew the display to an end. 'Thank you very much for coming,' I found myself saying once again, as a wave of relief washed over me. It might have gone a little better, but it could certainly have gone a lot worse.

Debbie returned from the far end of the field and gave me a comforting look. Various supporters from the village came over and said 'Very good' and 'Well done'. Debbie chatted with them, and a couple of children fussed over Swift, but I was still at a loss to know what the real response had been. The only group of visitors – the family in the car – had headed off almost immediately.

It was Gary who came over to speak to me last.

'Truly extraordinary, how Greg brought that ewe back,' he said, referring to what I had considered the low point of the display. 'Just amazing what you can do with those dogs, amazing.'

There was no doubting the sincerity of his words.

'Well, that's kind of you to say so, but to judge by the size of the audience tonight there aren't many people who share your appreciation.'

'No, you *must* keep it up. I reckon you've got the makings of a good evening's show there,' he said, a sudden forcefulness in his voice. 'Really, you *must.*'

As the last car drove out of the field Nick and Laura busied themselves jumping from one bale to another, while Clare headed home to shut the ducks in for the night. Debbie came over and we sat together on the end bale, surveying the scene. The sun was drifting down into the haze on the horizon, and the sky had a faint orange glow.

'Well?' she said.

'Gary seemed to have genuinely enjoyed it,' was the most optimistic response I could muster.

'I think everyone enjoyed it,' she said. 'The people I spoke to all sounded enthusiastic.'

'I think if we do it again, we've got to improve the advertising,' I said. 'It's not going to be worth it with those sort of numbers.'

'Well, we didn't exactly overdo it this time, did we?' Debbie pointed out.

By now Nick had stopped jumping on the bales and wandered over.

'Dad, why didn't lots and lots of people come to see your sheepdogs?'

Out of the mouths of babes. Before I could answer, Debbie had jumped up, wrapped an arm around him and begun leading him away towards the house.

'They probably will next time, Nick,' she said.

CHAPTER SIXTEEN
The End of the Gang

The moment I set eyes on the three hundred sheep gathered at the back of the dingy, unstrawed shed, I knew I'd made a big mistake.

The phone call had come a few evenings earlier. The voice had a Home Counties accent but was gruff to the point of rudeness. He hadn't even given his name, just the location of his farm, a fair distance away from our normal shearing round, and an outline of the size of the job he needed doing. Perhaps it was the paltry return we'd made from the first sheepdog display; perhaps I'd begun to see myself as Chris's replacement in organising the shearing and saw the opportunity to add to our shrinking list of customers; or perhaps it was just that I didn't know how to say no, but for some inexplicable reason I'd just said yes without thinking.

The first bad omen came when I called Geoff with the 'good news' that I'd found a new shearing customer with a five-hundred-strong flock.

Chris had already said that he couldn't manage any 'extra' sheep this year, so I was dependent upon Geoff for help, but when I'd called him his wife Marie had answered.

'Sorry, David, he can't come to the phone, I'm afraid,' she explained. 'His back went, shearing a few sheep for a friend of his father. He's flat out in bed, and can't even sit up. I'll have to get him to the doctor's somehow.'

I knew that Geoff had been suffering with his back this season, but we all did from time to time. He wasn't the sort to give in to illness easily, so I knew that if he was bedridden it was serious.

As I'd wished him a speedy recovery and put the phone down, I'd begun to wish that I hadn't been so hasty in my decision to accept the job. Now, as I surveyed this motley collection of livestock, that feeling was deepening by the minute. Perhaps the local shearers know something I don't know, I thought to myself.

I'd been there fully five minutes before a man ambled across from the farmhouse, clicking the heels of his dealer boots noisily on the concrete. As he came closer I could see his face was set in an aggressive-looking scowl.

'Morning,' I said, emulating his rather grumpy persona.

'You come to shear my sheep?' was his only response.

'That's right, I didn't catch your name.'

'Roy,' he said impatiently, extending a hand with all the relish of a man about to put his arm into a threshing machine.

This was clearly no time for small talk so I fetched the gear from the Land Rover and began setting up in the best-lit corner of the barn. As I sorted out my stand and put on my shearing moccasins, Roy loitered in an awkward silence. It was a relief to pull the cord on the motor, and make a start on what promised to be a long day.

I couldn't help thinking of the contrast between today and the previous job. A fortnight or so earlier, I'd been in good company, shearing fat ewes on a hot day. The twelve hours had flown by. Today, alone, as I tried to force the clippers through the tough belly wool of the first sheep of the day, I began to remember just how miserable shearing can be. Not only were the sheep in awful condition, thin and dirty, but they also lacked any of the grease in the wool that can make the job so simple. Worse still, as the day wore on they became progressively more soaked in the pools of dung lying on the shed floor. By mid-morning I too was coated in the stinking slurry.

Ordinarily the shearer concentrates on the shearing alone. The customer and whatever help he has to hand focus on filling the pens with sheep ready to be shorn, and on rolling the wool. It helps the shearer keep his rhythm going, and makes for an easier, shorter day. Roy seemed oblivious to any such etiquette and was nowhere to be seen. I mentioned it when I took a quick tea break halfway through the day.

'Oh, I've got other things on. Tell you what, I'll pay you an extra ten pence a sheep for doing all that as well,' he said.

Given the amount of extra work I'd been given, it was no surprise that by the time I packed up at seven that evening I hadn't even got through a third of the five hundred sheep he wanted shearing. By now I was thoroughly soaked in slurry.

'Can you turn them out for the night, and scrape out the shed for the morning,' I asked as I loaded the last of the equipment into the Land Rover. Roy looked at me as if I had demanded the unthinkable.

'I'm not turning them out to get them back in again, they can stay there until you've finished,' he said dismissively.

As I bumped back along the rutted drive, I took no notice of what was a glorious early summer evening. Good sense would have told me to cut my losses and not return, but I'd already done a gruelling day's shearing and I knew I wouldn't get paid if I didn't finish the job. I also felt some responsibility towards the sheep, who badly needed to be shorn. If I didn't turn up to shear them, I was sure that they would be left in that stinking shed until Roy could find a replacement.

Arriving back at the farm the following morning I found the sheep in an even more pitiful state, their bellies dripping in slurry, looking hollow and forlorn. Roy seemed oblivious to their plight, but I decided I'd seen enough.

'I'm not shearing them till they've been out at grass for a few hours,' I said. After a brief argument, Roy reluctantly agreed, and let the flock stream across the yard to where they pulled at the grass as if they'd never eaten before. The suspicion that this was a man who shouldn't be allowed to keep livestock was growing within me.

It was almost two hours before I finally started on the first sheep of the day. At least by then the ewes had been given a chance to fill their bellies a little, even if their cleanliness had hardly improved.

Roy spent most of the morning walking to and fro across the yard, shouting grumpily at sheepdogs, sheep and anyone else who came within earshot, so I was more than mildly surprised when, as one o'clock arrived, he suddenly turned into a paragon of hospitality.

'Come in and have some lunch,' he said.

The house was tatty inside and all but bare. It was a scant

sixties re-creation of the way a Devon farmhouse should have looked. His wife, whom I had seen fleetingly in the distance, sat at the table reading a magazine. She made no attempt to get up. Roy unscrewed the top of a new bottle of whisky and slopped out a half glass. With a gesture that I took to mean that he was offering me a swig from the bottle, he downed the glass without pausing.

'Right, let's get on.' He headed back for the door, without a further word to either me or his spouse. Clearly he had stock that was waiting to be sworn at.

The fact that I worked my way through the rest of this part of his flock that afternoon owed less to my shearing technique and more to my determination to get away from this place for good. Roy had told me the rest of the sheep were kept elsewhere and would be available for shearing tomorrow. The thought of finishing them off and giving this miserable man his bill spurred me on.

The sun was little more than an ember glow below the horizon when I arrived home, tired and thoroughly cheesed off. There were still jobs to be done, but fortunately Debbie had spent the afternoon out with the dogs, checking on the sheep that I hadn't had time to see before I had left in the morning. Collapsing into a chair, I phoned for news of Geoff. Deep down a part of me had hoped for a miracle, but any idea of Geoff rising Lazarus-like from his bed was quickly dispelled.

'He still can't move,' Marie said, sounding rather worried. 'They're going to take him to hospital in an ambulance tomorrow.'

As I hauled myself out of bed the following morning, the prospect of a third day's shearing at Roy's farm had filled me with dread. The thought of returning to his flock of neglected geriatrics was bad enough, but two days of 'tough' sheep is far more painful on the body than shearing good ones and it had been a real struggle to stand upright, so I'd been quietly relieved when the weather came to my aid. There had been a light spring shower overnight. Knowing his sheep would need a couple of hours to dry out, I got on with my routine around the house, beginning with tackling the morning's post, in particular a fat brown envelope marked with the telltale red stamp: DEFRA.

'A little more light reading,' Debbie said with a sarcastic twitch of an eyebrow as I shook loose what looked like half a ream of paper on to the kitchen table.

'If I read everything that came through the post, I'd never have time to leave the house,' I replied, spreading out the wodge of pages.

The heading of the letter accompanying the enclosed booklet was 'Rules for the disposal of dead stock'. I groaned and stuffed the letter and booklet back into the envelope, then added it to the pile of similar literature on the table with a sigh. I was pouring myself a fresh cup of tea when I was struck by the strange silence in the kitchen.

'Where are the children? They're a bit late, aren't they?' I asked Debbie.

'With the puppies. You know they won't do anything in the mornings until they've fed them.'

The children had been besotted by Gail's litter and had insisted on keeping them in the house during their early days. Debbie had managed to cope for three weeks, but the combination of pups breaking free from the confines of their box, making unpleasant messes in ever more unexpected places, and testing their milk teeth on the furniture in the living room had proven too much, and a week ago they had been moved outside.

I looked out of the window towards the shed in the yard

where the pups now resided, but saw no sign of movement. However, as I stuck my head out of the kitchen door to scan the other side of the house, it wasn't hard to spot them. Nick was approaching with a pup clenched determinedly on to his trouser legs, and a second pup hanging on to the tail of the first. The girls were a little way behind, with a flurry of pups lolloping and stumbling around them. Gail was also in tow, but was obviously trying to keep a distance from her offspring. Every time she paused, a ravenous gang nuzzled under her belly, trying desperately to slurp the last of her milk. Given the razor sharpness of the youngsters' teeth at this stage in their development, there was no chance of her hanging around.

All eight pups had now been named, not with the standard sort of sheepdog names, but with an abstract selection of titles that the children considered best described their appearance or character.

'Here, Elvis,' I could hear Laura calling towards one of the stragglers. Elvis had apparently been shaking a little when they'd come up with his name. I had named the runt of the litter, a long-legged little character who looked rather like a cross between a whippet and a vole. I'd christened him after the latter.

Vole was towards the rear of the mêlée of puppies as they arrived in the house, playing, chewing, biting and growling as they went. Nick picked up one of the dog pups and carried it, legs dangling, under his arm towards me.

'Dad, can we keep Eddie?' he said, his best pleading angel face fixed in place.

'Not that one, Dad. This one's the sweetest,' Laura said, clutching a ball of fluff to her chin.

There could be no winning in this situation. I'd committed to keeping one pup at Borough Farm in the short term, but whichever one I chose, I would be in trouble with one or other of the children, and likely as not all three.

Truth be told, I'd already taken a shine to a bright little bitch pup that the children had called Mist. Typically, despite my subtle interventions, none of the children had so far decided this was the one to keep. Once more I tried to introduce a little logic to proceedings.

'We can't keep a puppy just because it's the sweetest, Laura! You've got to work out which one is going to be the best sheepdog.'

The look I got from all three of them was a mixture of disdain and dismay.

'Breakfast, you lot, you'll be late for school,' Debbie interrupted, placing a pile of toast on the kitchen table.

As the children sat down to breakfast, the pups rolled around on the kitchen floor, biting at the occasional toe that dangled enticingly close by. Mist, on the other hand, had found another plaything.

'Oh no, Dad, look at that!' Clare lowered the toast from her mouth, a look of disgust on her face. Mist had somehow poached the thick brown envelope that had fallen from the table and had dragged it across the kitchen floor. She had now turned tail and was squatting over it, creating a large, steaming pool that was spreading over the distinctive, red DEFRA stamp.

'Now there's a sign of intelligence; this must be the one we keep,' I said, sweeping the pup swiftly away, then giving her a ruffle behind the ear as I shut the door behind her.

———

The passing of the warm spring rain had left the farm smelling sweet and fresh. I couldn't bring myself to rush away on such a fine morning, so before heading off shearing, I ran a flock into the handling yards, shedding off some lambs ready for marking. The thought of having the first batch ready for market lifted my mood.

By 10.30 I could put off the inevitable no longer. With a freshly printed invoice in hand, I set off to meet Roy and – I hoped – complete the third and final round of his wretched shearing job.

I'd agreed to meet him at some 'off ground' that he rented, which turned out to be a sheltered valley several miles from his farm. It was soon obvious it was not a place he visited too often. As I wearily set up the shearing gear once more, Roy started to gather his flock. They arrived in dribs and drabs in the pen we'd set up, but in one glance I could see that half a dozen of the sheep were suffering from one of the most unpleasant problems that afflict sheep: fly-strike.

Fly-strike crops up during the late spring and summer, when blowflies crawl into the wool of sheep where they then lay their eggs. When these larvae hatch, they set about making a meal of their living host, their burrowing – and the smell it produces – attracting more flies to lay more eggs until the poor animal is literally eaten alive by maggots.

No self-respecting shepherd should allow things ever to reach this stage. Modern veterinary science has produced an array of preventative dips and sprays, all of which are highly effective. It was blatantly obvious that Roy hadn't treated his sheep. Nor indeed had he even bothered to check on them during this dangerous time of the year. As more sheep arrived into the pen, I saw more and more victims, some alive with maggots. I'd reached breaking point by now and couldn't hold back.

'These sheep are plastered in maggots,' I protested, when Roy closed the gate behind the last sheep. 'Why on earth haven't you done something about them?'

For the first time, I felt I'd struck a nerve, perhaps even pricked his conscience. He looked uneasy as he searched for an excuse. It didn't take him long to find a scapegoat – or scapesheep, to be precise.

'Haven't had the chance to get over here much,' he said, 'and besides, they all get in the brambles once they've got the maggots, you never find them in there.'

I was appalled at a farmer putting the blame on to his sheep like this.

'But you can treat them to stop them attracting the flies,' I said. 'It's only basic shepherding.'

By now, however, I realised that basic shepherding was something Roy knew next to nothing about. Having briefly been on the back foot, he was soon back to his arrogant self.

'Just shear them off,' he said dismissively. 'They'll be fine.'

I didn't have any choice. I owed it to the sheep to take their wool off. That would at least get rid of the maggots, and prevent re-infestation until the wool re-grew, but the open sores and wounds where the larvae had been would need medication urgently.

'Have you any antibiotic spray?' I asked Roy. 'A lot of these sheep will need treating before you let them go.'

He wandered off and rummaged around in the back of his truck before eventually emerging with a solitary can of the relevant spray. I started up my machine and picked out one of the worst affected sheep.

Shearing a maggoty sheep is nothing short of vile. As the clippers run along the sheep's skin, they cut through a heaving layer of squirming larvae. Other areas of the sheep are raw and open, with wounds so deep that it defies logic that the animal can survive.

As I finished off the first sheep, Roy appeared at my side

and began applying the treatment to the poor creature's many wounds. It was far too little, far far too late.

I was virtually speechless by the time I arrived back that evening. I collapsed in the chair once more and tried to forget the awful day's work. I'd left Roy by pressing my bill into his hand and he'd promised to pay it within the week. I'd rather have taken the money there and then, but by the end I was just happy to get away.

The news at home did little to lift my mood.

'I spoke to Marie earlier,' Debbie said, producing a life-saving cup of tea. 'They took Geoff straight into surgery and operated on his back. He had spinal fluid leaking and pressing on to his spinal cord.'

'Is he going to be all right?'

'They think so, he's feeling better already, but they're not sure if he'll be able to do manual work again, let alone any shearing.'

I felt terrible at having so obviously underestimated the seriousness of the situation. This was a massive blow to someone who had worked in farming all his life, and would have huge implications for him, Marie and their family.

It was going to affect us as well. Even at the current, pitiful rates, shearing a few thousand sheep a year made a significant contribution to the farm accounts. In the darkest days, it had been this and Debbie's part-time work as a nurse that had kept us afloat. Now that was gone.

The conversation inevitably drifted back towards the sheepdog displays.

'You never know, if we get the advertising right, then in time we might be able to earn as much from the displays as from the shearing,' Debbie said. 'I was talking to someone at the tourist information office, and they reckon visitor numbers are going to be well up this year.'

I wasn't sure whether she was just trying to inject some optimism into what had been a thoroughly depressing day. Inwardly, I found it hard to share her view of the glass being half full, but I tried my best to seem positive.

'Well, let's hope that you're right.' The idea that the cash earned from the shearing round would magically be replaced by holding a few sheepdog demonstrations on the farm still seemed complete pie in the sky to me.

The kitchen had a rather subdued atmosphere over breakfast. The three children sat, cross-legged in a row on the floor with their backs to the Aga, their toast and cereal forgotten as they sat quietly cradling the selection of puppies that they had already retrieved from the old stable.

'Dad, you're so cruel,' Laura said, gently stroking one as she watched me making myself a mug of tea. 'They aren't nearly big enough to leave home yet.'

Laura, Clare and Nick knew that this was the day the first of the litter left home, so they now sat fussing over their favourites. A friend, Tony, was coming to choose one of the pups, to take to the farm where he worked in Oxfordshire.

The puppies' inevitable departure for new homes always came too soon as far as the children were concerned. Gail's litter was now eleven weeks old and could easily have been sold three weeks earlier. They would have been settling into their new homes by now, if it hadn't been for the relentless pressure of the children to let them stay 'just a bit longer'.

Their attachment was perfectly understandable. There can be few sweeter-looking creatures than collie puppies. Yet those chocolate-box qualities are also the reason so many working dogs end up in completely unsuitable homes – and, ultimately, rescue centres. People are drawn to the collie's good looks without realising that, as a breed, the working collie just can't cope with the confines of a domestic life. It needs the physical and mental stimulation of working sheep that is hard-wired into its system. Which was why I always insisted on selling my puppies to working sheep farmers.

The longer the pups were on the farm, the harder it was going to be for the children to say goodbye. That, however, was not a message I was going to get across today.

'Poor Gail, when she realises you've sold one of her puppies she'll probably never work for you again,' Laura said.

I knew my protests would be in vain but I had to try to defend myself.

'Gail's probably forgotten that they are even hers by now,' I said, failing abjectly to elicit any interest let alone sympathy from the children. 'She hasn't been feeding them for weeks.'

I might as well have been talking to the wall.

'He definitely can't have this one,' Clare said, half demanding, half pleading, as the puppy she was holding pulled away from her lap and ran across the kitchen floor. 'He's my favourite . . . ouch!' The puppy had returned and dug its teeth into one of her toes, then started to pull at her sock, growling as it played.

All three of the children would have carried on having fun with their boisterous playmates all morning if it hadn't been for the sudden appearance of a rather peeved-looking Debbie.

'Please can you get ready for school,' she said, pointedly looking at the clock, and meeting their protests with an implacable gaze. 'It's only one of them leaving today,' she reassured them. 'The rest will still be here when you get back this afternoon.'

Reluctantly Clare and Nick headed out of the door, calling the pups with them as they went. The last to haul herself up was Laura. As she headed off to brush her hair and get her book bag ready, she gave me one final, sulky look. Today I could do nothing right.

––––––

The cacophony that precedes the morning feeding of four hundred sheep had not long died down when Tony arrived. The dogs heard his van first and barked their usual noisy welcome. Tony and I went back fifteen years, to my early days as a shepherd in Kent. Watching him amble into the shed, to me it all seemed a lifetime ago now, but the memories were happy ones nevertheless.

We had worked together on a large estate in south Kent owned since the death of her husband by a rather 'grand' lady called Mrs Ellis. Mrs Ellis was something of a throwback to another era, the pre-War days of maids and manservants, when life was divided between upstairs and downstairs.

It had been Mrs Ellis's late husband who had laid the foundations for the life to which she had become accustomed. He had moved to the Romney Marsh in the forties, when he had become one of the first farmers to plough it into vast, fertile fields and plant crops. By the 1980s, when I arrived there, its wheat and barley were the basis of a thriving and successful business that also included a large flock of sheep.

Mrs Ellis's imposing house was a testimony to the success they'd had, and two gardeners were employed to keep the grounds in order. As the shepherd, though, I had little contact with events at the main house – but there were a few occasions on which our paths crossed.

Sheep have a peculiar gift for heading in precisely the direction where they are not wanted. It's as if they sense our anxieties and decide to tweak them even higher, so my flock would take every opportunity to make for Mrs Ellis's well-manicured garden. There were numerous occasions when they would inexplicably veer off course to make an attempt on a rose bed or geranium patch. Either that or individual sheep would escape from a mob, only to reappear among the newly flowering wallflowers. Most such forays were, by luck or judgment, thwarted, but inevitably a few were successful. I remember being summoned by the lady of the house at six one morning to deal with half a dozen ewes in the garden. They had spent the night chewing bedding plants, fertilising the drive and, worse still, upsetting the stone bird-table outside her kitchen window.

Mrs Ellis considered the day-to-day problems of the farm beneath her. She left issues like collapsing lamb prices, disease outbreaks and underproducing ewes during the lambing season to her farm manager. The pillaging of her precious garden was another matter entirely.

'You had better remove them immediately,' she had told me haughtily.

On that occasion I managed to save the day with a grovelling apology, accompanied by a bunch of flowers. The flowers, I might add, were *not* picked from her garden.

Not all the farm's catastrophes were down to my delinquency, thank goodness. On another occasion the early morning phone call I received from Mrs Ellis was informing me that a heron had chosen to help itself to breakfast from her garden pond which was stocked with beautifully marked koi carp.

'I'd like you to dispatch the bird,' she said with a firmness that was faintly chilling.

By the time that I arrived, however, the bird had flown away, which was fortunate. If she had discovered that I'd turned up at the house without bringing any cartridges for the shotgun, she'd probably have sacked me on the spot.

It was, however, a memory that involved Tony that brought the biggest smile to my face.

At ten to eight each morning during the lambing season, Mrs Ellis would invite Tony and me to the house for a full cooked breakfast. This was an admirable gesture. Quite apart from the work involved in preparing a three-course breakfast, she had to put up with the overpowering smell we carried around with us during this hectic time of the year – a blend of dung, sweat and amniotic fluid. Not pleasant.

Conversation was often a little strained: Mrs Ellis was not a lady who enjoyed smalltalk, at least with the likes of Tony and me. I'd learned that questions about the lambing required nothing more than a 'Fine thank you, Mrs Ellis' in response.

It was Tony who overstepped the mark. I'd tried to get the conversation moving a little by asking Mrs Ellis what she thought of the previous day's Budget.

There was a long, slightly chilly silence.

'I shan't answer that,' she responded eventually. 'Instead I shall ask how was the Budget for you?'

The obvious response was to point out that she had never paid me enough to qualify for tax, but I thought better of it.

'To be honest, I haven't had much chance to hear what was in it,' I said diplomatically.

'And how was the Budget for you, Anthony?' she asked, turning her attention to the second shepherd at her table.

Tony was not exactly a man to observe the finer details of etiquette. He had rolled up his sleeves, revealing his tattooed arms, and was more interested in wolfing down his breakfast, and in particular his fried eggs. They were a little on the runny side and he was compensating for their tendency to slip through his fork by lowering his chin to plate level.

'Don't interest me, to be honest with you,' he said, without lifting his chin. 'No one never gives me nothing.'

Mrs Ellis's voice was edged with anger. 'No one gives you anything in this life, Tony. Everything that you have has to be earned!' she declared, before storming from the room. It was just as well that she departed. Tony and I would have choked on our bacon if we hadn't been allowed to burst out laughing.

———

I gave Tony a quick tour of Borough Farm, and as he cast a knowledgeable eye over the sheep in the shed, I felt slightly embarrassed by my rather modest set-up. Until recently he'd been working on a large estate near Swindon, a well-equipped farm with up-to-date buildings and equipment. However, the last twelve months had transformed his life too. Most of the estate's income came from shooting and assorted other non-farming leisure activities, and foot and mouth had put a temporary halt to visitors to the estate. It had been the final straw for Tony's employer. The fifteen hundred sheep had barely provided a profit in recent years, particularly once the wages of his shepherd had been taken into account, so Tony's boss had wasted no time in putting the entire flock under the auctioneer's gavel. It was a hammer blow for Tony and his young family too. Not only was his livelihood under threat, but so too was his house, which was tied with the job.

But Tony was a grafter and his hard work had not gone unnoticed. A near neighbour had offered him a job, looking after his vast flock of sheep, and had agreed to pay the rent on the tied house.

The downside of this arrangement was that Tony had found himself tending to two thousand breeding ewes and five thousand fattening lambs, spread over a vast swathe of Oxfordshire. Working thirteen hours a day seven days a week, Tony was barely seeing his children.

'They have to come with me at the weekends; they're getting good at chasing lambs through a footbath,' he said with a stoical grin.

Tony had five dogs of his own, but, just like me, he was faced with a couple of his better dogs growing older, and he was finding it well nigh impossible to do his job.

'Trouble is, you can wear all the dogs out, with flocks as big as I'm running,' he told me. 'You try and put a thousand lambs into a pen in one mob.' So it was that he'd come to me for a puppy from Gail's litter.

It was lunchtime before we got around to the serious business of inspecting the pups. When we had bred Ernie a few years earlier, it had been obvious which member of the litter to choose. Ernie was the only pup to jump from his run and find his way to the house. When he had also gone to the sheep field and chased sheep at twelve weeks old, he had as good as assured himself of a place in the kennel. This litter was now nearly eleven weeks old, but as far as I could see not one stood out from the rest. Everyone has their own way of picking a pup. Some choose the boldest, some the most timid as it might be the easiest to train. Others go for the runt of the litter, working on the assumption that it has already learned to stick up for itself so will be more prepared to face up to a belligerent ram later in life. I didn't expect Tony to opt for this selection method somehow.

There was only one scientific method of selection that I had come across: a series of tests performed on pups at forty-two days of age that would point to the puppy's character later in life. It sounded great. However, it didn't seem to incorporate a test to tell if the dog would be an expert at working sheep, which rendered it all but useless as far as I was concerned. As a result, I sometimes used a method that was far from scientific. If there was no discernible difference between any of the pups, I asked the children to choose.

It turned out Tony's method was just as imprecise. The nine puppies had spread themselves across the yard, just outside their stable. They were well dispersed, playing in twos or with noses to the ground in search of chewable items.

'I'll have that one over there,' Tony said, pointing at one that had found itself a small piece of wool, and was now running full speed around the yard with its highly prized possession. 'See it's already got a nose for sheep, or wool at least.'

To my immense relief, I noticed that the puppy was not one of those that Clare and Laura had been making such a fuss of before they went to school. To my mild disappointment, neither was it Vole.

'Are you sure about that?' I said. 'You could have that one if you like.' I pointed to the small lanky pup with brown legs, who was a little apart from the rest.

'No, I'll let you keep him,' he said with the hint of a smile 'They say the ugly ones turn out the best, so he should be a real cracker!'

Tony had only just turned the corner at the top of the drive, when the children arrived back from school. As Clare made a bee-line for the stable, it was Laura who began the cross-examination of their 'cruel' father.

'Which one did he take?' she asked, anxiety written all over her face.

'It's OK, he didn't fancy any of your favourites,' I said, content in the knowledge that I'd steered him away from any of the dogs I'd seen in the kitchen that morning.

The look of relief on Laura's face was short-lived, however. As we approached the stable door, Clare had re-emerged, a look of admonishment on her face.

'Dad, I don't believe it, you let him have Johnny, he was my favourite.'

As I'd suspected, I was going to do nothing right today . . .

CHAPTER SEVENTEEN

Elusive Customers

The sight of a Land Rover pulling into the yard induced a sudden wave of anxiety. I knew immediately who the driver was, even though I'd not met him before. Ivan ran a large farming business up on Dartmoor, a sprawling operation which included a large herd of beef cattle and a substantial flock of sheep. I'd been anticipating his arrival for a few days now, as I'd agreed to take in two of his dogs, Trim and her kennel mate Taffy, for a couple of months, during which time I'd train them to be working sheepdogs.

The good news was that it represented another way of

supplementing our income, albeit a modest one, but as I watched Ivan clambering out into the afternoon sunshine I wondered at what price. I nursed a faint feeling of dread, mainly because it had been the ever-mischievous Andrew who had paved their way.

'You'll probably be getting a phone call from Ivan,' he'd announced one evening a week or two earlier, visibly relishing the look of surprise that immediately spread across my face. 'I've told him you'll train a couple of dogs for him. Won't be a problem, will it?'

Experience had taught me to treat Andrew's every word with suspicion.

'Why aren't you going to have them?' I asked, knowing he'd trained dogs himself in the past and had previously sold Ivan a very good sheepdog.

'You'll see,' he said, with his usual verbal economy.

Now, as I approached the back of Ivan's Land Rover, I had the feeling I was about to discover precisely what he meant. I peered in through the grille to find two dogs pressing themselves forward.

'Shall we let them out and have a look at them?' I said.

Ivan looked at me, his cloth cap shielding his eyes from the glint of the sun, mild concern etched on his gently lined face.

'I'll have to watch Trim a bit,' he said, as he carefully lifted the back grille and inserted his arm, feeling for the collar of the nearest dog. 'She might not come back if she . . . oh bugger!' Mid-sentence and with lightning speed Trim had twisted from Ivan's grip and squeezed her head through the narrow opening of the rear door. She dropped to the ground, then dashed, whippet-like across the yard.

'Trim, 'ere Trim, 'ere, good girl,' Ivan shouted. The dog took no notice. Nose to the ground she ran to the first gate, squeezed under it, and found the freedom of a large field. In what appeared to be only half a dozen bounds she was at the far end, her mainly white coat little more than a dot against the far hedge. As we watched, she stopped, apparently found a gap, pushed through it and disappeared up into the shave of woodland above, that ran along the old railway line.

Ivan stood there looking disconsolate. 'Oh bugger. I don't rightly know how we'll catch her now,' he said, shaking his head slowly. He paused for a moment, screwing up his eyes as he scanned the horizon. 'I wonder where she'll go,' he said, before adding, almost as an afterthought, 'Don't worry though, she won't chase the sheep.'

That last comment was, I supposed, meant to reassure me, and, in the short term at least, it did. The downside was that it wasn't going to help in the coming weeks. It was rare indeed to come across a Border Collie that had no interest at all in sheep, and if that was the case with Trim, there would be little chance of my converting her into a working dog.

Ivan and I watched in silence for a couple of minutes. Only the stirring of a small herd of cattle in a neighbour's field nearly a mile away, gave an indication that she might still be in the vicinity. I could see Ivan was anxious.

'I don't rightly know what she'll do now,' he said, looking rather concerned. I tried to reassure him.

'She had her nose to the ground as she was going. She may well follow her own scent and retrace her tracks, once she realises she's on her own,' I said, trying at the same time to sound like the expert on sheepdogs I was being paid to be.

Rather than continuing with what seemed like a fairly pointless guessing game, I suggested we turned out attention to Taffy. He had sat patiently throughout the drama with Trim, uttering not so much as a whimper from the back of the Land Rover. That was something, I thought to myself. At least he wasn't a hot-head like his partner.

Taffy was, to put it mildly, a rather unusual-looking collie. Thickset, and just a little overweight, he had a thick coat that was mainly black, but a mottled white around his collar. His large ears were laid back across his head and he had one bright blue 'wall-eye' which gave him a slightly unnerving appearance.

'Let's get him into his kennel, before we go looking for Trim,' I suggested as Ivan lowered the tailboard. Taffy it seemed was more compliant than his partner. He sat next to the Land Rover straightaway, gently wagging his tail, panting slightly.

'How old is he?' I asked Ivan, noticing a yellowness of the teeth that I wouldn't have expected to see in a young dog still waiting to learn his craft.

'Coming on four,' said Ivan, seemingly oblivious to the fact that at this age a dog should be at the peak of its working life. We walked him across the yard to the kennels that I had built in preparation for my new arrivals. Once more Ivan looked slightly dubious.

'Have you a bit of mesh that we could put over as a roof? He can be a bit of a Houdini.' It had never occurred to me that my new kennels weren't completely dog proof. With six-foot high mesh sides around the run and a cosy kennel at the back, I had thought them ideal accommodation. Ivan, however, was adamant, so we began tying a collection of old gates and mesh across the top of the run.

We were just finishing off when Debbie appeared from the house, sounding slightly puzzled: 'Jan next door has just been on the phone. There's a strange dog in their barn, but she says it doesn't look like one of ours.'

Ivan looked as relieved as I felt.

'I thought she'd turn up if we left her long enough,' he said with a slight ring of triumphalism in his voice.

'Has she shut her in?' I asked.

'I think so. Is it one of Ivan's then?' Debbie asked, looking rather bemused. 'How has she got over there from here in the last twenty minutes?'

'It's a long story. I'll pop over there and pick her up.' Debbie started to walk away, still looking a little concerned.

'I hope you're not expecting me to walk this dog for you,' she said, turning back towards me as she headed into the house.

I assumed Ivan was going to accompany me, but was soon put straight.

'I couldn't leave you to pick her up, could I? I've promised to call in and see some cows for a man in Bideford on the way home,' he said. Without even waiting for my response – a rather resigned 'No problem' – he was on his way back to the Land Rover.

'I'll give you a ring to let you know how they're doing, in a couple of weeks,' I called as he disappeared up the drive.

———

An hour or so later, having made my apologies to Jan, I returned Trim to the scene of her crime, firmly attached to a piece of rope. Working sheep with her would have to remain a distant dream for the time being. For now, my objective was to try to win her trust. So with the lead lengthened to twenty or so feet, I took her for a walk around the field from which she had escaped earlier in the day. I called her back to me at regular intervals, pulling gently on the rope to force her to my side, and making an enormous fuss of her once she had obliged.

I was just beginning to think we were bonding rather nicely when a flash of mottled black rocketed past us. Taffy it appeared had lived up to his Houdini reputation and come to join us. Unfortunately, he had now spotted a dozen ewes at the lower end of the field.

'Taffy!' I bawled as the black flash headed in their direction, but to no avail. With the terrifying speed and precision of a guided missile, he zeroed in on his target, and in particular on a sheep at the left of the group. As impact occurred the poor ewe was sent sprawling.

'Taffy, no!' I shouted as I sprinted after him.

To my amazement, as I arrived within twenty yards of him he stopped dead still and lay looking up at me, wearing an expression that I'm sure was meant as an apology.

After a moment or two he started to skulk back towards me, but as he did so, the running sheep took his eye once more. In an instant he switched again to guided missile mode, off this time in pursuit of the left-hand sheep, which he soon intercepted with all the precision he'd shown seconds earlier.

'Taffy, noooooooooo!'

Once more he dropped to the ground and lay looking in my direction.

I approached him with more care this time. He lay flat, with the same pathetic naughty schoolboy look, his tail twitching intermittently as I bent down to secure his lead.

'You, Taffy,' I growled at him. 'We've got some work to do with you.'

He looked up at me and wagged his tail once more. If he could have spoken, I'm sure that he would have told me that he couldn't help it.

A few minutes later and the two dogs were back in their respective runs. I spent another ten minutes tying up any gaps through which Taffy might squeeze himself, and headed indoors, but even before I could kick off my boots at the back door the sound of Trim barking excitedly had me haring back across the yard.

The scene that greeted me was barely believable. Taffy had climbed on to the top of his kennel, and had leapt up to grab the mesh that I had tied over the roof of his run. He was now dangling there, hanging by his teeth like a fish on a hook, frantically scrabbling with his front paws, trying to get a grip on the gap that he had created at the join in the mesh.

'Taffy!' I bawled. 'What are you up to now?'

The response was quick. He loosened his jaws and dropped to the ground like a lead balloon. Then – yet again – he looked up at me with the same ragamuffin grin I'd seen in the field. I was to see that look a lot in the next few weeks.

––––––

I was sitting at the desk in the office, putting off the filing by flicking through the newly delivered edition of the local farming journal when Debbie appeared.

'Someone on the phone about a puppy,' she said with a hopeful look.

Four of the litter had been 'booked' before they had been born, and word of mouth had brought one further buyer to the door. It was pleasing, not least because each of them had said they were looking for a 'well-bred' puppy. Most of the buyers had been local farmers, and once more the voice on the end of the line had a strong Devonian burr to it. It was also very abrupt.

'So how much are you asking for theses 'ere pups?' the caller asked, dispensing with the usual pleasantries. The price of a good working sheepdog pup has never borne any relationship to their true worth as far as I am concerned. Fully trained dogs go for reasonably large sums of money – £3,000 or more at the big sheepdog sales, like Sennybridge in mid Wales – but even this I would argue scarcely reflects the time, hard work and skill that goes into training these animals. Sheepdog puppies, however, have always struggled to fetch a reasonable price, certainly in comparison with other breeds.

I've always found it somewhat unbelievable that a dalmatian pup at eight weeks is worth five or six times the value of a well-bred sheepdog of the same age. Just because it has spots. So, as with the rest of the litter, I was determined to get a fair price for what I knew was potentially a very good working dog.

'These pups come from an excellent lineage. Pedigrees as good as any. Their parents are good working dogs on and off a trials field. I'm asking £120 each,' I said, sounding as businesslike as I could. To my mind it was relatively cheap, the equivalent of employing one man for two days. To the caller, however, it was akin to daylight robbery.

'What? I never paid more than £20 for a dog before; always had good ones, mind,' he said, his irritation obvious.

Where have I heard that before? I thought to myself.

'I'll have to think about it. P'raps you could ring me in a week or so,' he said curtly.

'Will do, if it's still here,' I replied, trying my best to remain the calm salesman. 'Give me your name and number and I'll call you at the end of next week.'

It turned out the man's name was Duckworth, a sheep farmer from the other side of Barnstaple. I duly noted his name and number, but quietly hoped I'd find another home for the puppies rather than have to call him back.

———

As I looked at my watch a sense of déjà vu washed over me. It was nudging towards twenty to six, and once more I was staring at the top of the lane, willing some of the passing cars to turn left into the fields of Windcutter Down.

After long deliberation, Debbie and I had finally decided we would persevere, and hold a sheepdog demonstration every Wednesday evening over the summer. If we gave it our best shot through the peak weeks of July and August, by the start of

September we'd know for sure whether it was a venture that was worth pursuing.

A big factor in our decision was the groundswell of support that seemed to be developing in the local community. The Woolacombe tourist information office had been actively promoting the event, as had the Heritage Centre Museum in Mortehoe. Both thought the displays would appeal to the area's visitors, especially in the evenings.

Encouraged by their confidence, I'd invested in a rather more professional-looking sign at the end of the drive, but it had been Debbie who had been the real driving force. For the last week she had been busying herself touring every local campsite, hotel and guest house, delivering posters and bundles of the fliers we'd hastily made up. Now, on the first Wednesday in July, it was time to wait and – for me at least – to worry.

'If no one turns up this week, then I think we might as well give up,' I'd said, to a look of faint scorn from Debbie as she arrived in the field just after 5.30. Within moments, however, the first of what turned out to be a stream of cars turned into the field. By the time my watch showed a couple of minutes to six there were nearly sixty people positioned on the bales – and barely a familiar, local face among them.

The most surprising discovery was that addressing sixty people was no more daunting than addressing the dozen and a half I'd talked to back at the end of May. Indeed, to judge by the ripple of acknowledgment that greeted even my introductory 'Good evening and welcome to Borough Farm', the applause seemed to be easier to come by, and there were fewer of those deafening hushes that followed my attempts at humour. The reaction remained warm throughout the display that followed, and each of the dogs performed well – even Ernie, who for once listened in part to what he was being told.

As I brought the hour-long display to an end there seemed a genuine appreciation in the crowd's reaction, and a gaggle of children fussed over Swift, much to her delight.

'That went well,' Debbie said, as the last of the cars left the field a few minutes later.

'Suppose so,' I said reluctantly, 'but it's hardly going to be our salvation if the best we can manage is fifty people.' I must have sounded like the prophet of doom. Debbie, however, was in far more optimistic mood.

'It's a big improvement on last time. This time everyone was a holidaymaker – and did you see the way the children swarmed all over Swift at the end? Come on, cheer up.'

Deep down, I still remained unconvinced.

———

Debbie cupped her hand over the receiver and giggled at me.

'It's that Mr Duckworth,' she said.

'Oh no, not again,' I whispered, being careful to cover the mouthpiece as I took the phone.

In the fortnight since I'd first spoken to him, Mr Duckworth had become something of a running joke between Debbie and me. A week on from his first enquiry I'd called him back, as promised. It had been a disappointment that I'd even had to. I'd placed one of the two remaining puppies with a farmer from Exmoor, but still had the last one, a nice-looking dog the children had christened 'Whiteface' for obvious reasons. Reluctantly I'd found the note I'd scribbled with Mr Duckworth's name and given him a ring. His reaction was unexpected, to say the least. It appeared he had no recollection whatsoever of our previous conversation.

'Puppy, why would I want a puppy?' he said. 'No, I don't need another dog. I've already got one here.'

'But you called me about it.'

'No, no, you've got the wrong chap.'

I'd put down the phone and shaken my head in amused disbelief, and had never expected to hear from him again. Now here he was.

'I been thinking,' he began, this time in a rather self-important voice. 'If you've still got that pup,' there was a meaningful pause, 'and if you's prepared to be negotiable on the price, seeing how there's only one left, then I might be interested.'

By now I was beginning to feel that I'd invested enough time in Mr Duckworth – but I did want to sell the pup. I also, if I'm honest, found dealing with him very amusing.

'Well, why don't you come down and see the pup, see the parents working, and then we'll discuss the price,' I said.

'Ah, you still got 'im then? 'Aven't been able to sell 'im,' he swiftly interjected, as if spotting a chink in my armour. There was a pause for a moment while he decided which approach to take next. 'I'll have a think about it,' he said. I hung up the phone convinced that there was no chance whatsoever I'd see him.

The sun was setting when the sound of crunching gravel drew my attention to the gleaming new BMW pulling into the drive. The man that stepped out was smartly dressed in jacket and tie. The only hint as to his identity was a cloth cap. I nearly fell over when he introduced himself. For some reason in my own mind I had decided that Mr Duckworth was a struggling small farmer. I'd put his reluctance to pay my asking price for the pup down to a simple inability to afford such an extravagance. I'd clearly been well wide of the mark.

It is generally reckoned that the best way to sell a puppy is to introduce a potential owner to the dog's parents. It's the perfect way to show what the puppy could potentially become when it grows up. So I took Gail out into the fields to demonstrate to Mr Duckworth the lineage that he had the chance to buy into.

He didn't say much as we made our way down to the steeply sloping field in front of the house with Gail ahead of us. There was a small flock of ewes at the bottom, and I set Gail off to gather them. She disappeared down over the steep bank, and out of our view like a dog on a mission. A minute or more passed and Mr Duckworth turned to me, clearly unimpressed.

'What's happened to this so called . . .' But before he could finish his sentence, the sheep appeared over the hill in front of us, Gail tucked in behind. Suddenly Mr Duckworth's tone changed.

'That's incredible,' he said. 'In all my years of farming I never seen a dog work like it.'

Encouraged by his response I worked Gail a little longer, and Mr Duckworth's praise became, if anything, even more fulsome.

'I'll tell you, I know every farm this side of North Molton and no one's got dogs as good as that,' he said.

As we walked back to the yard to have a look at the pup, he was still eulogising about what he'd seen. 'So this 'ere pup is going to be as good a dog as his mother, is he?'

'With the right training and handling, there's no reason why not,' I said honestly.

When I led him into the kennel, he was no less enthusiastic about the pup.

'What a beautiful pup, that's just what I'm looking for. Long coated, and nicely marked.'

We talked for a minute or two more, before I moved in to close what I felt was a certain sale.

'So you'd like to take him, would you, Mr Duckworth?' I said.

There was another pause, during which he took off his cap and replaced it several times. A pained expression spread across his face as he did so.

'I'll have to think about it,' he said, without a hint of embarrassment in his voice, and with that he made off to his car.

I tried to suppress my laughter as he drove off up the lane. How much thinking does a man need to do, I asked myself as I headed to the kitchen.

'If Mr Duckworth ever rings again, be sure to tell him I'm not in,' I told Debbie.

CHAPTER EIGHTEEN
An Off Day

As I took shelter from the late morning sun under the shade of an oak tree, the nerves were once more getting the better of me. The faint knot in the pit of my stomach had been there since first light and had rid me of any appetite for breakfast. I knew there was no one to blame. After all, no one had forced me to enter the English National Sheepdog Trials. No one had made me drive halfway across the country against my will. What's more, I kept telling myself, if I was a true sportsman I should be able to deal with a little attack of butterflies. By now, however, I realised I could talk to myself until I was blue in the face. Nothing was going to make me feel any better.

The setting for this year's event was the grounds of Knaresborough House in Yorkshire. I'd driven up from home overnight and had arrived at three in the morning to find the grounds covered in a thick, low mist. I'd swung the Land Rover into a suitable parking spot and snoozed for a while. It was only when I got up at around seven that I saw I'd almost driven straight into one of the lakes that dotted the grounds. Perhaps it was a good omen, and things were going to go well, even better than last time, I told Greg and Swift as we all shook out the stiffness of the long night.

Like the singles competition, the brace requires a high standard of ability and obedience from the working dogs who compete, but it also relies heavily on the instinctive partnership between the two dogs. Greg and Swift had developed this over years spent working together. They had an understanding of each other's abilities, and an acceptance that each partner had his or her own role to play, but their second place two years before at Powderham Castle had been about much more than this. To achieve anything like the perfection needed to compete at this level, even the best dogs need a huge amount of training and polishing. Their success had been the result of long hours honing the finer points of the sheepdog's art. This time, however, I'd found it harder to make the time for practice, and, as the National drew nearer, I'd been thwarted by the fact that Swift had come 'in season' and had had to be separated from all the male members of the pack, including Greg, while she remained that way. Last-minute polishing had become an impossibility.

Which was another reason why I was feeling so queasy this morning. Once the competition was under way, however, I knew I would have to pull myself together. As I left the shade of the tree and walked the specially prepared grounds, I breathed in and tried to banish any negative thoughts.

Several hundred cars and vans lined the perimeter of the course. Many had their backs open, to let their canine occupants cool down. Some even had purpose-built trailers with ventilated sides to keep their star performers in the best possible environment before competing. In the middle of the ground stood a row of white marquees, at the centre of which was the home of the International Sheepdog Society, custodians of the sport for over a hundred years, and organisers of the three days of competition. At the edge of the course stood the judges' box. It would be the occupants of this who would decide our fate once more. If they were anything like as demanding as the audience, they were going to be a tough bunch to please, I thought to myself as I took my place among the large crowd.

A few yards from me a pair of grey-haired men watched sceptically, muttering to one another about the dog and handler now on the field.

'Don't know what he's doing here with a dog like that,' snorted one of them, adjusting his cloth cap with one hand, while tapping the ground in frustration with the crook that he held in the other.

'If you stamped a foot at him, he'd turn tail and run. No good for a day's work on the hills,' said another portly chap, leaning forward in his chair.

To me, the handler hadn't made any obvious blunders up to now, but as I looked more closely I saw that his dog was having limited success in penning his five sheep. Every time they got anywhere near the pen, one ewe was leading the rest around the side. The dog was giving ground and this old ewe knew it. His eye had conveyed a weakness and now, having lost the respect of the sheep, he was losing control completely. The old man was probably right.

''Tis a sad day when the sport of working shepherds is lowered by a dog that couldn't do a day's work,' he muttered, as if responding to my thoughts.

This was the critical audience before which I was going to have to perform. So when the announcer proclaimed the end of the morning's singles runs, adding 'Will the brace competitors now stand by?', I felt the knot in my stomach tightening even more.

The brace is a more specialised discipline than the singles so – unlike the individual contest, in which one hundred and fifty compete – it's limited to nine pairs in all. Three of them run at lunchtime on each of the three days of the event. We were the last of the trio to go today.

I wanted to avoid giving Greg and Swift an overly long wait at the competitor's gate, so I figured that I would watch the first run before getting them from the car.

The first competitor in the brace was soon reminding everyone why only nine people in the country were prepared to try to run two dogs at a time. Brace trialling can, when it goes wrong, go completely wrong, and that was what was happening here. Naturally, the two elderly men near me had plenty to say about it. I tried to blot out their voices, for fear of turning myself into a nervous wreck, but hard as I tried, I still caught the phrase 'the whole sport's in decline' drifting in my direction.

Better make a good job of this, I thought to myself as I walked back to get the dogs.

I had left the Land Rover sheltered from the sun under a blanket that was almost completely covering the rear windows. Greg and Swift had had a good run earlier in the morning, and now, after a good rest, should have been perfectly prepared for their lunchtime trial, but as I drew back the blanket I discovered Greg had decided on an entirely different kind of pre-match preparation.

With time to kill and the boss absent, Greg had chosen this moment to make one last entry in the Sheepdog Society studbook. The two of them now stood 'tied' in this bizarre act of canine entanglement. I felt the colour draining from my face.

The competitor before me had by now walked on to the field. If his run went well, he might be out there for another twenty minutes, but if it went badly I would only have a fraction of that time to get ready. Any thought of victory went out of the window. Now my only ambition was to survive the next half hour. I couldn't withdraw at the last minute; my only hope was that Greg could!

As I waited for the dogs to separate, I kept half an eye on the events on the field. Already the sheep had been returned to the handler and were setting off on the drive. A cold sweat descended over me.

Finally, after what seemed like the longest ten minutes ever, I was able to make my way across to the competitor's gate, hoping no one would notice Swift's slightly distracted demeanour. Greg didn't have a care in the world. His tail wagged constantly, he looked thoroughly pleased with life. Swift, on the other hand, had her head down and was panting heavily. Hardly a team focused on the job in hand, I thought.

'Oh, 'ere he is,' came a call from another competitor – a man that I recognised from the competition two years earlier. Back then there had been too many entrants and he'd tried to persuade me to step down in his favour. He clearly still bore a grudge. 'You reckon you can do it again this year, do you?' he said in a slightly condescending tone.

'We'll see,' I said, determined that this was not the time to start making excuses.

As I made the short walk to the post the announcer introduced us to the audience, reminding them of our second place at the last English National. I could almost feel the weight of the audience's anticipation bearing down on me.

'Now you two, please try to remember what you came here for,' I whispered.

At the opposite end of the long flat field a packet of ten sheep appeared. It was a job to estimate the distance, but they must have been at least four hundred yards away. Greg was suddenly alive to the task, scanning the horizon. He spotted the sheep and waited for a command. This was encouraging, but Swift was more of a concern. She lay at my right, looking completely nonplussed.

Swift needed to be sent first, as her outrun was the longest. I gave her the command to go, but, instead of launching herself towards the sheep at breakneck speed, she ran twenty yards before dropping to the ground, panting. Greg, on the other hand, was clearly invigorated by his recent activities. He didn't wait for his command and was already off at full speed, bang on course, and likely to arrive at the sheep long before Swift, thus ensuring the judge would soon be docking points fast.

I wasn't sure whether Swift intended to take part at all now. I whistled to her, and she wearily climbed to her feet. As she picked her way lackadaisically among the trees that lined the course, I gave her a couple of shrill blasts of the whistle, just to remind her of where she was meant to be going. After what seemed like an eternity, she reached a position behind the ten sheep, where Greg was already waiting.

Any relief that I felt at this point was soon quashed, as, for some unknown reason, Swift overshot the spot where she should have stopped running by so much that she turned the sheep back towards Greg. In trialling terms this was becoming a disaster. The ewes were now pointing back towards the pen from which they had just been released.

Although another whistle sent Swift scampering back into position, the damage was done. All ten sheep had now decided that they were going to resist any attempt to turn them in my direction. A murmur from the watching audience suddenly reminded me of the two old men. I could only begin to imagine what they were saying about me. "E's got two dogs and they can't even fetch the sheep between them.'

The situation was only being made worse by the course, which was long and almost completely flat, making it practically impossible to see where the dogs were in relation to the sheep. I made an educated guess and blew a few whistles, with a distinct ring of panic. I knew that everyone would now be watching, and not for the right reasons. The entire audience was probably wondering if they were about to witness the ultimate of trialling humiliations – not just one but two dogs failing to lift their sheep. It had probably never been seen before.

Just then there was some relief. Slowly, very slowly, after what seemed to have been an age, the first sheep turned. Now, as far as I could make out, they had started down the course towards me. The two dogs I could now see tucked in at either flank; for the first time I felt as if I was in some sort of control.

The sheep drifted off to the left of the course, and seemed to be missing the first set of gates. A whistle to Greg slowed him down, and a flank to Swift sent her to her right. I could feel a little tension lift as the sheep passed through the hurdles.

A few more steadying whistles and the sheep were back at my feet. The next task was to turn them behind me and start on the first leg of the 'driving away' element of the run. After a few moments of defiance the ewes turned, but as they drove away I could sense from Greg that he wasn't in the mood to listen. I blew a 'stop' but he took no notice. I growled at him, but there was still no reaction, he just kept pushing on. 'Nothing for it,' I decided. 'Just have to make sure that Swift keeps up.'

Keeping things moving at a steady speed is a prerequisite for a successful run, but with Swift up and alongside Greg, and both of them now pushing on too fast, the sheep broke into a trot.

The first set of drive gates loomed up ahead. The sheep changed gear from a trot to a run, and veered to the right. Swift ran to turn them in once more, but Greg wanted to take control. Seeing the gates just ahead, and guessing what his next command might be, he took it upon himself to whip around to the left, anticipating the start of the cross drive. However, he had moved far too soon. Swift was still out of her correct position and, although I brought her back around as quickly as I could, the result was a complete mess: dogs and sheep at stalemate in the middle of a field, with none of them knowing which way to go next. There could be no saving of pride from here on. No matter how good the rest of the run, there would be no reassuring 'bad luck' and a pat on the back as I walked from the field. The whole run was becoming a fiasco.

By the time the sheep eventually entered the shedding ring, I felt beaten and dejected. To make matters worse, when I finally managed to shed the sheep into the required two groups of five, the course director appeared.

'Judges want you to shed again,' he said without pity. 'One of the sheep was outside the ring.'

That was enough. I waved to the judges to signal my retirement and called the dogs over. Together we headed from the course, deliberately choosing the route that kept us furthest from the spectators and the old men in particular. I really didn't want to hear their comments.

I didn't feel like staying through to the end of competition, but somehow I spent the afternoon watching events from a distance, trying to avoid the knowing smiles that inevitably came my way. The contrast to two years before could hardly have been greater. Then I had left with the feeling of having excelled my wildest expectations. By five in the afternoon today I'd had enough. I gave the dogs one last walk, and tried to slip away quietly.

'Didn't have the luck this year then, did you?' came a gloating voice that I recognised in an instant. It was the same character who'd given me grief for being late for my run.

'Swift's been in season and . . .' I said, then immediately realised I'd made a mistake.

'That shouldn't make any difference,' he snapped. 'I've run a bitch in season many a time, and it's never made any difference to her. Mind you, I've got proper working dogs.'

With that he marched off, clearly content that he'd made his point. I watched him go, struck dumb by the wretched man's arrogance.

'Don't worry about it, Greg,' I said minutes later as he thudded his tail against the side window of the Land Rover and we bumped our way out of the grounds of Knaresborough House. 'We all have our off days. I just wish you hadn't had yours in front of quite so many people.'

Within a few days I'd begun to see the funny side of my all-too-public humiliation. I didn't have much choice, such was the mileage Andrew was making out of my misfortune at the

National. I'd spoken to him a couple of times since the debacle and he'd laughed so much his ribs must have hurt. In the end I'd decided to join in. Even so, as Wednesday and the latest demonstration day arrived, the thought of running Greg and Swift as a pair in front of an audience again filled me with doubts for the first time. Why would anyone pay to watch my dogs run on this sort of form?

I was about to head up to Windcutter Down to prepare for the evening's demonstration when the postman appeared. In among the usual collection of bills and circulars, the scrawled handwriting on a large brown envelope caught my eye. It was addressed simply 'To The Shepherd' and had an incomplete address. Inside, sprawled in crooked lines across the page, was a letter clearly written by a very young hand. It read: 'Me and my brother really liked seeing Swift rounding up the sheep and we liked meeting Swift at the end as well. From Philippa and Christopher.'

If the letter had put a smile on my face, the picture attached to it almost brought a tear to my eye. Carefully carried out in pencil, the drawing was of a square-bodied dog with matchstick legs, presumably Swift, and two other four-legged animals, presumably a pair of sheep. Flanking these three was a two-legged figure which I assumed to be me. I had swirly eyes, and in my outstretched hands held a crook and a microphone.

The accompanying letter from their mother explained that Philippa had enjoyed the display so much they had insisted on writing to me – and Swift, of course. She said that Christopher had been so impressed he wanted to spend his life in the fields. 'The only thing is, he can't decide whether he wants to be a shepherd or a sheepdog,' she joked.

When Debbie appeared in the kitchen from upstairs I passed her the letter. The effect was identical, a smile then a slight wobble of the chin and an 'aaaaah' as she looked at the picture.

'You see,' she said, 'I told you people were enjoying themselves. Now you've had your first piece of fan mail.'

'Well, I think it was aimed at Swift rather than me,' I said, mildly embarrassed at the very notion.

Whatever the truth, I set off for Windcutter Down with a new enthusiasm.

———

I was in the sheep pens when Debbie appeared late in the afternoon.

'You're never going to believe this,' she said, a big grin on her face. 'I've just had Mrs Duckworth on the phone.'

'What – Mrs Duckworth, as in Mr Duckworth?' I asked, slightly puzzled.

'Yes, that Mrs Duckworth. She was asking if we still had the last puppy?'

It was almost a week since I'd had my visit from Mr Duckworth. During that time he'd become even more of a joke among the family. Laura and Nick had once called me to the phone saying it was a 'Mr Woolworth'. I'd picked up the receiver only to hear the line was dead. The children thought it was so funny they'd tried the same trick again half a dozen times, but with less success. As it happened, more pressing events had pushed the selling of the puppy to the back of my mind, and it was still available.

'She said they want to come and buy it,' Debbie went on. I gave her a doubtful look. 'No, they want to come and buy it now. They're on their way over. She says she can't wait until this evening.'

We looked at each other in disbelief.

Sure enough, half an hour or so later Mr Duckworth's BMW

glided into the yard. It was clear who really wore the trousers in his household. Mrs Duckworth said they were prepared to pay the asking price of £120. Her husband dutifully peeled off the notes.

Usually, part of me was concerned that the dog may not be going to a suitable home, but I had a feeling the Duckworths would actually be a rather good pair of owners. Mrs Duckworth helped confirm this by asking about his feeding, vaccinations and where he was used to sleeping.

'The grandchildren will be over at the weekend,' she said. 'Dare's say they'll be wanting him in the house.'

'I'll give you a call in a week or two, let you know how he's getting on,' said Mr Duckworth, as they climbed back into the car, Mrs Duckworth clutching the pup on her lap in the front seat.

As I watched the last of the litter leaving the farm, I turned to Debbie. 'He'll probably be asking for a refund if it hasn't learned to drive the tractor and shear the sheep by the time it's six months old!'

CHAPTER NINETEEN
A Killer on the Loose

The low mist of the early morning was slowly lifting. Out on Morte Point, the gentle gush of the receding tide was revealing a landscape of glistening rocks, deep blue pools and shillet-filled gullies, but for once the almost surreal beauty of the place was lost on me.

Forty feet or so below me lay a ewe spread-eagled on the rocks, and very dead.

I was already fairly certain this was no accident or act of suicide but the work of an uncontrolled dog. To begin with, the ewe had fallen to its death at a point on the coastal path, close to the village, where the cliff edge forms two sides of a triangle. At this time of the year this was a busy spot, a walkway used by countless numbers of visitors. On top of that, this was the second time I'd had to recover a sheep this week. This usually only happened two or maybe three times a year, often on a false alarm, but two evenings earlier I'd had a phone call asking me to rescue a sheep stuck on the rocks only a couple of hundred yards from where I now stood. On that occasion I'd succeeded in catching her and dragging her to the grass above. She'd run off unharmed. The ewe below hadn't been so lucky, and recovering her presented a very different challenge.

As I made my way to the Land Rover to fetch a length of rope, I felt a familiar mixture of emotions. I had understood the dilemma that came with renting the National Trust land at Morte Point from the outset. Increasingly, the countryside – and sites of outstanding natural beauty like this in particular – has come to be seen as a place of recreation for both the local and urban populations. Indeed, such have been the changes in the way the countryside works, it would be easy to argue that this is its primary purpose.

You only have to look at the village of Mortehoe. At the beginning of the last century more than half of its men were dependent on farming for part of their living. The situation could not be more different now. Only four farms in the parish are still working, and between them they employ only two part-time workers. The tourism industry, on the other hand, provides work for people in their hundreds and, in financial terms, for every £1 generated by farming in the parish, £500 is earned through visitors. The foot and mouth crisis had illustrated how much the Woolacombe area had come to rely on them. As with the rest of the country, all of the footpaths had been shut down for many months. The drain on the area's income was huge. Everyone – not just the farmers – felt its impact.

So, striking the balance between the needs of the tourist and those of farmers like me was a difficult task. Yet the fact remains that the Point is also a workplace. To me it beggared belief that some visitors could be so totally oblivious to the interests of people whose livelihood depends on the countryside that they let a dog run amok among a flock of sheep. I simply couldn't afford to lose valuable members of my flock like this.

I returned to the cliff with the rope tied to the back of the Land Rover, and threw the loose end over. I then walked a little way to a spot where it was easy to climb down on to the rocks. It wasn't long before I had reached the ewe, and had begun looping the rope around her legs. She could, I thought, really have stayed where she was. Nature has its own way of recycling, and the local gulls were already making the most of the chance of a non-fish diet, but I knew that wasn't feasible. Debbie had already been fending off numerous calls from locals and visitors, so I had no option but to remove it.

Back up on the cliff top a few minutes later, I pushed the Land Rover into gear. I felt the rope tightening and pressed forward. Behind me the sheep's carcass rose unceremoniously from the rocks. A few minutes after that I loaded the stinking cargo into the trailer.

As I cast my eyes around the headland, part of me hoped I didn't spot the culprit. Such was my anger at whoever it was that had allowed their dog to drive the ewe over the edge, I couldn't have been sure of my reaction. Farmers are still entitled to shoot dogs found worrying sheep, but it is a rarity to be forced to take such drastic measures any more. Apart from everything else, we understand dogs too well. They all have an inbuilt desire to chase sheep; it is part of their genetic make-up. If they have met stock before and have shown they can behave, then I have no objection to dogs being left to roam free, but otherwise, owners have to exercise caution and common sense and keep their dogs on leads.

As I made my way out through the village and homewards, however, more mundane thoughts began to seep into my mind: all this had lost me a good chunk of a working day that could have been put to much better use. Summer has always had its list of chores to keep me busy, jobs like worming and docking the lambs and running them through the footbath, but with new strings attached to my bow, the demands of July and August were greater than ever.

———

Deep down I still wasn't sure which of the new ventures were worthwhile, if indeed any of them were. The Wednesday sheepdog demonstrations were looking the most promising. Audience numbers were unspectacular, but for the last few weeks at least, they had been high enough to suggest that – one day – they might provide a regular source of good income. They were certainly more worthwhile than selling a few puppies, which had proven far more eventful than Debbie or I had imagined. There was no question in my mind, however, that training Taffy and Trim was bottom of the list in terms of potential profitability. Neither was progressing at the sort of speed that justified the money Ivan was shelling out on a weekly basis, and, in truth, I'd begun to find it a real effort to work with 'the away dogs' as I called them. I'd far rather spend time with Jake. Today, once more, I had to force myself to take them to the training field.

Since they'd arrived on the farm, Trim had made the better progress of the two. She'd responded well to the lavish praise I'd given her and come to respect me as a boss worth returning to. As a result, I no longer feared letting her run free. That said, however, she was still a sheepdog of limited potential and I was only teaching her the most basic elements of sheepdog work.

That morning I wanted to practise her outrun, the most fundamental task of a sheepdog, but one that can be difficult to teach. I asked her to wait by the gateway into the field while I took up a position closer to the sheep. It is a simple technique that should entice the dog to run in a wide circle around a scattered flock.

With a simple 'Away, Trim' she was off, running effortlessly with greyhound-like strides towards her prey. Her good work ended when she arrived at the sheep. Trim's problem was not the over exuberance that you might expect from a dog in training; in fact her failing was quite the opposite. When she came face to face with sheep she exuded a general air of ambivalence – and so it proved again today. As soon as she arrived, the sheep moved off in front of her, but, instead of flicking her eyes from one to the next, weighing up their every movement as a good sheepdog should, she seemed to stare right through them.

I tried a couple of commands: a left and a right. Trim trotted in each direction without much purpose, and certainly without breaking into the extraordinary pace that I knew she had.

I called her back to me, and her tail twitched with recognition as she approached. I rubbed her nose. She was certainly a vast improvement on the dog that had been delivered, but she was a mile from being the sort of dog with which you could gather sheep on the open expanses of Dartmoor. I clipped her on a lead and turned my attention back to her partner.

I'd learned my lesson with Taffy and had now taken to tying him to a tree root with a heavy, chain lead while I worked with

Trim. He'd been watching the proceedings patiently enough and, as I approached, his ears tucked back and he seemed to say that butter wouldn't melt in his mouth. However, over the last few weeks I'd learned not to trust this look.

'Are you going to behave today?' I asked him as we walked across the field, rather closer to the sheep than I had started with Trim.

He couldn't answer but, with the husky-like way he was pulling on the lead, I drew my own conclusions. He was ready to have what I'd come to call one of his 'missile moments'.

During the one phone call that I'd had with Ivan since I took on his dogs, he had made a significant confession.

'Taffy's got a habit of chasing one sheep off by itself,' he said. 'I've tried all sorts, but I just can't get him to stop it.'

During the past couple of weeks it had become clear this was a major understatement. On more than one occasion, normally when I'd just begun to think we were getting somewhere with his training, he would suddenly explode, detaching an outside sheep from the rest of the flock and chasing it across the field, pulling mouthfuls of wool as it went. By the time I arrived to remonstrate with him, he would be lying, ears tucked back, with a look of complete apology. He was incorrigible.

As I got ready to work him today, I couldn't help thinking back to the incident out on Morte Point. The dog or dogs that had caused the problems were displaying all of the raw, untrained instincts that Taffy had. If for no other reason than this, it was important I persevered in trying to break this habit.

To be fair to Taffy, he had shown some progress in the past week. For perhaps the first time, I felt less on edge as I walked backwards across the field, encouraging him to push the sheep towards me. Each time that he showed signs that he might break into missile mode, I gave him a firm 'Lie down, Taffy' and, to his credit, on most occasions he did. After fifteen minutes or so of this he'd gone through the session without once picking off a single sheep. He didn't have much of an idea about how to be a sheepdog, though, and, at the age of four, it wasn't going to be easy to teach him anything at all, but I knew if I could keep this up I'd feel justified in charging Ivan his fee, at least. It cheered me up a little.

I arrived back in the yard to be greeted by the sight of Nick, breathless and bursting to tell me something. In the past months he had perfected the art of wearing the face to suit the occasion, and at this precise moment it was his best, long, serious one.

'Dad . . . Dad, bad news, I'm afraid,' he said. At first I wasn't too concerned. Bad news in Nick's book could just be that he had pulled a carrot from his own personal patch in the garden to discover that it had been attacked by the slugs. On this occasion though, he was right to look so grave. 'A man's just phoned and says that there's a sheep swimming in the sea.'

What it lacked in detail, it more than made up for in impact. I headed back to the house, a familiar mix of dread and resignation welling up inside me. In the kitchen Debbie looked irritated.

'It was the Swansea coastguard,' she explained. 'They've had a report of a dog chasing sheep off the cliffs – somewhere just to the west of Bull Point, nothing more specific than that. The chap who phoned them said that one of the sheep was swimming out to sea.'

I was incredulous that this could be happening again. With so many people walking along the North Devon coast path, there are bound to be a few incidents of this kind, but now I'd had three in the space of eight days.

Clare's ears had pricked up at the news.

'Can I come?' she said, reaching for her wellies. This was obviously a job far too exciting to miss, and by the time I had reached the Land Rover, Laura had joined us as well.

––––––

We arrived to discover a dozen people gathered at the landward end of Wide Gut, to the west of Morte wells. At one hundred and eighty feet, it is the deepest inlet on this section of the coast. The cliff sides, rugged and spectacular, fall away as they reach the mouth. Looking downwards to the rocks here, I could see the focus of the crowd's attention. A lamb was perched uneasily on a rock right on the edge of the gully, and as we watched the swell of the sea sent a wave surging through the inlet. For a moment the lamb looked as if it must be swept from his refuge, before the wave subsided, leaving him sodden, but still rooted upon the rock.

'Can you save it?' asked one of the onlookers.

'I should think so, but I'll have to be quick,' I replied.

Parking the Land Rover among the bracken, I collected the rope from the back and slung it over my shoulder. Clare and Laura were already ahead, slipping and climbing down the fern-covered bank that ran alongside the gully, heading towards the spot beneath which the lamb was stranded. The short tussocks of grass met the gully with an abrupt drop. The girls lay on their tummies, and the three of us peered over the edge. The lamb's legs were now submerged by the relentless waves. If it felt any fear there was no sign of it, just a look of resignation and misery. From the top end of the gully it had seemed like a tricky climb down to the lamb, but as we peered down from this new vantage point, I saw the rocks on which it stood were reasonably accessible: just a short climb, and a bit of a scramble. I dropped the rope down, just in case the lamb proved difficult to lift back up.

'What do you want us to do?' Laura asked as I started my descent.

'You've got to dive in and save me if I fall,' I replied.

Laura looked serious for a second before a lively grin spread across her face.

'It won't matter if you fall in, as long as you don't let the lamb fall in as well,' she said with a laugh. That sounded ominously like the start of a second animal rights activist in the family. Clare had long ago decided that the needs of her animals far outweighed those of any mere human.

Thankfully I found the climb down was as easy as it had looked. A minute or two later, I was face to face with a heap of woollen misery. He must have been in a state of shock as he made little attempt to avoid capture. I slipped a loop in the end of the rope around his middle.

A shove, a heave and a lot of lifting, and the lamb was back on dry land. I sat down with the girls, hoping they might have been impressed by their father's heroics.

'Just as well you didn't drop it,' said Clare, glancing up towards the rim of the inlet high above us. 'Those people up there wouldn't have been very impressed.'

The onlookers were still visible on the skyline although it was equally clear they were dispersing now the entertainment was over.

'And nor would I for that matter,' Laura chipped in, the grin spreading once more across her face.

The lamb on the rocks may have been saved, but that evening it became apparent that it had not been the lone victim of this latest attack. The caller to the coastguard had described a sheep 'swimming out to sea'. Sure enough, that evening I got a phone call to tell me a dead lamb had washed up on the shingle cove of Rockham, only a few hundred yards away from where we'd found the first one. I made arrangements to collect it from the beach the following morning. As I lay awake in bed that night there was no escaping the obvious conclusion. Three incidents was way beyond coincidence. There was a dog, or maybe a pair of dogs, out of control on this section of the coast.

———

The days of early August should be the warmest of the year, but as Debbie and I looked westwards towards the sea on the first Wednesday of the month, the outlook was anything but summery. Instead, the sky was a deep, gloom-laden grey. According to traditional country lore, cows are the key to predicting the weather: if they lie down rain is imminent. In my experience, however, it's the birds that can really sense the weather. The swallows were 'pipping' noisily around the outbuildings, rather than flying high as usual at this time of the afternoon. The air was heavy, and they could obviously sense that rain was on its way.

'I wish it would just get on with it,' Debbie said, the frustration obvious in her voice. 'At least then we'd know to cancel.'

I understood how she felt. The prospect of abandoning what should have been the fifth sheepdog display of the year was hard to take, particularly given that we'd seen a steady stream of cars the previous Wednesday evening. The healthy turnout had been due in large part to Debbie's hard work delivering a stack of 'fliers' to every holiday destination that we could think of in the area. This week, however, her efforts were going to be in vain it seemed.

The first rain arrived, not as the fine droplets of a gentle drizzle, but as spots big enough to fill a glass. As they splattered on to the concrete, we ran for the house.

'We'll leave it for twenty minutes and see what happens,' I

said, pulling the door closed. 'If it carries on like this no one will turn up anyway.' The drumming of the rain on the windows was interrupted by the ringing of the phone. 'That'll be someone asking whether the display's on,' I said as Debbie picked it up.

The concerned look that spread across her face immediately told me it was something more serious.

'Swansea coastguard . . . again,' she said, handing over the receiver.

Minutes later I was going through the all-too-familiar ritual of loading a rope and a dog into the Land Rover. By the time I headed up the lane the rain was teeming down. With any prospect of a sheepdog display now abandoned, I gave Debbie a lift to the top of the drive and left her there in her raincoat, in position to turn away any hardy souls who turned up.

The location I'd been given by the coastguard was further along the northeast coast, near where the dead lamb had been washed up two days before. This section of coast is separated from the rest of Morte Point by the Uscott Valley which means access is impossible by normal roads. Instead I would have to head through the ground of a neighbour, Ray.

I found him in his yard, clad in yellow waterproofs, at the wheel of a tractor hauling a trailer containing the rubbish bins from the campsite he ran with his wife, their daughter and son-in-law Brian. Ray was one of the few locals who could truly claim to have been born and bred in the village. Now in his seventies, he was a goldmine of stories from Mortehoe's past. For many of his farming years Ray's sheep had grazed Morte Point and the adjoining coast of Bull Point, so he understood the problems I'd been having better than most. Indeed, Ray had faced situations far more testing than that facing me now. He'd farmed cattle as well as sheep, and had more than once had to rescue cows from the rocks. He still had some cine footage of one that had fallen fifty feet on to the rocks at Uscott. The poor animal was winched to safety by a team of coastguards, police and locals, having suffered no more than a broken tooth. So the problem I described to him seemed nothing out of the ordinary.

'I don't mind going down on the rope to get it,' he said almost eagerly when I explained what the coastguard had told me.

'Let me have a look first,' I said. 'You know what these reports of stuck sheep are like; it probably won't even need any help.'

According to the coastguard the sheep was on an inaccessible ledge. I knew that the cliffs along this section were quite severe, and I didn't fancy the prospect much. I fancied it even less when I arrived at the spot.

The sheep, or to be more precise, the lamb, didn't take much finding. It was perched on a heather-covered ledge about two-thirds of the way up the cliff, thirty feet from the top, and fifty or sixty feet from the sea below. Unlike the lamb stranded on the rocks a couple of days earlier, this one didn't seem to have a care in the world. It lay quietly chewing the cud, with its eyes half shut. Tucked under the overhang, it was cosy and dry as the rain was passing clean over the top of him.

I wouldn't say that I suffer from vertigo, but neither would I be a natural steeplejack. Looking down, I felt a tingling spread from my knees upwards. A gust of wind caught my back as the rain rattled on the hood of my coat. I gulped hard, stood back and tried to think of a plan. Swift had joined me from the Land Rover and had spotted the sheep. I called her away from the cliff edge. I certainly didn't want her deciding to mount her own rescue.

The lamb had to be saved and, as the shepherd responsible

for it, I felt it had to be me who brought it back up, but I could see already I wouldn't be able to do so immediately. The tide was high and the froth turbulent around the base of the cliff. I'd have to wait until the sea had receded.

On the way back through the yard, I found Ray. 'I'll not be doing anything until the tide's out,' I told him.

'You'll have to wait 'til mid morning: low tide's around ten,' Ray said. 'I'll come and give you a hand if you like. I've got all the gear.'

A few harnesses and the like wouldn't go amiss, I thought to myself. 'Fine,' I said before heading homeward. 'See you around ten tomorrow.'

Over breakfast the next morning Nick was fit to burst with 'what ifs'.

'What if the lamb falls . . . what if you fall . . . what if someone else's dog comes along and chases all the sheep off the cliff?' This last question seemed particularly apt just at the moment. Clare had a more practical question, however.

'What if you phone the Air Sea Rescue,' she asked. 'The helicopter might come and pick it up for a bit of practice.'

It wasn't a bad idea. The Air Sea Rescue was based only a few miles down the coast, and they frequently trained along the cliffs. However, the prospect of them sending me a bill for the operation soon put paid to the idea. It would probably cost more than the value of the entire flock.

'I don't know about that, but you lot could come and help,' I said. The response was enthusiastic. 'I'm heading down there now. Why don't you follow with your mum in a few minutes?'

Ray's wife Margaret waved me down as I drove through their yard once more.

THE DOGS OF WINDCUTTER DOWN

'Ray's already there,' she explained, a concerned look on her face. 'Don't let him go down on the rope, will you?' From the anxious smile she gave me as I drove on, I guessed that might prove more difficult than it sounded.

Sure enough, Ray was already at the cliff when I arrived.

'I've brought all the gear,' he said, pointing to a coil of blue rope. I looked around briefly for a harness, but quickly realised that there wasn't one. This was 'the gear'. The rope didn't even look long enough to reach the base of the cliff. I discreetly fetched my own from the back of the Land Rover, together with an iron spike.

While I hammered the long spike into the ground, Ray peered over the edge. He was clearly finding the idea of launching himself extremely tempting.

'Look, it's an easy climb from here,' he said, pointing to the thirty feet of near sheer loose rock that lay below us. 'I could be down there in a minute, then all I've got to do is to catch it up.' Ray was beginning to resemble a terrier who had just caught scent of a rabbit down a hole.

'No, Ray, really. I'll climb up from the bottom, it'll be easier that way. Could you sort the rope out from the top?' Ray looked disappointed, but reluctantly conceded that as it was my lamb it was my right to fall off the cliff. He secured the rope to the spike, then slung it over the edge. It seemed to take a long time to fall to the rocks below. The lamb stood up for a minute, but appeared unconcerned. I started the short walk down the side of the valley, to a point where a narrow path led from the pasture above. As I did so, Debbie, Clare, Laura and Nick arrived, the younger two running on ahead in their anxiety to be first on the scene.

The tide was well out as I picked my way across and down to the base of the cliff. I found the rope, and looked upwards. It's peculiar how the mind works. Looking from the top, this approach from the bottom had seemed by far the easier option. Now suddenly the climb down from the top looked as if it would have been less difficult.

I had little time to dwell on it, however. A small cluster of heads peered down from on high. I could make out Ray with the second rope, Debbie, the children and a few walkers who had joined them. All were looking expectant. This was no time for me to waver. Wrapping the rope around my middle I started upwards.

The previous night I'd decided that one side of the rock face looked slightly more inviting than the other. It was flatter and had what looked like ledges and cracks that might easily provide footholds. This short section of the coast is made up of Morte Slate, a thin and crumbly rock that is constantly eroding and can be dangerous to climb, but this side looked less likely to fall away when I climbed it. Most importantly it wasn't a sheer climb either.

As I pulled myself up on the rope, the route soon proved less straightforward than I'd thought. The footholds and ledges were either too far apart or I found that handfuls of solid rock fell off immediately I put any weight on them. Somehow, however, I hauled myself slowly upwards, making sure not to look down as I went. It wasn't pretty. A competent rock climber would certainly not have looked so clumsy. I stopped half a dozen times to tighten the rope around my waist, but eventually I grabbed at the root of some overhanging heather and pulled myself on to the ledge, fifty feet above the sea and ten from the lamb.

Any hope that the sheep would co-operate quickly evaporated. It pricked its ears, looked at me for a moment then, before I could pull myself to my feet, dived down over the opposite side of the ledge. I ignored it for a second; this was no place to start a chase. I straightened my back, gazed around, then downwards, and promptly tightened the rope around my middle again. I then pulled up the crook I'd tied to its end.

The onlookers were now only thirty or so feet above, close enough for me to see the look of slight concern on Debbie's face. Nick was struggling to keep still with all the excitement and even the girls seemed captivated by the events.

Ray's expression was rather different, however. Why are we waiting, it seemed to say.

The lamb had chosen to take refuge about ten feet below me, on a mound of loose-looking shillet. He was standing on an angle so acute that it seemed he must surely start to slide towards the drop below him. I waved my crook towards him, and he turned back towards the cliff face, jumping into a gully that ran down it.

I wasn't going to follow.

As I decided what to do next, I heard a shout from Ray and looked up to see him lowering the second rope, the one on which to pull the lamb up. Clearly he felt that by now the rescue should be all but over. I tied a loop in one end of this rope, and passed it back through itself. Plan B, I'll lasso the wretched animal, I thought.

Out on the prairies and pampas of North and South America, the lasso is an everyday tool of the stockman's trade. It's not so common in North Devon, however, and it soon showed. As what must have been my twentieth loop of rope failed to uncoil itself around the lamb, I found myself wondering whether, by any chance, the group of onlookers included a holidaying cowboy.

The sound of voices above made me look up. This time I saw Ray in conversation with his son-in-law. It was soon obvious that the pair of them had decided I needed help, and I was just calling up to Brian that I was OK when he grabbed the rope with one hand and launched himself over the edge of the cliff, abseiling towards me in the midst of a small avalanche of earth and rock, and at a speed that took me aback. Within a few seconds he stood by my side.

'He's down on all that loose rock,' I said as we weighed up the situation together. 'I reckon if I stand on that mound of shillet, the whole lot may go.'

Brian half muttered something that I didn't quite catch, then grabbed the rope with one hand once more, before scurrying off down the scree in hot pursuit. A shower of shillet sprayed beneath his feet, but the mound showed no sign of collapsing. For a moment the lamb thought about running further down the gully, but, when he realised it ended abruptly a few feet below, he came to a halt. Brian saw his opportunity and, still holding the rope with one hand, jumped behind the lamb before thrusting out his other hand to catch it. As he looked back up towards me, I threw the second rope down and he secured it around the lamb's middle. A minute or so of pulling from me and it was back at my feet. Brian soon scurried up behind. Now the rest of the family came into their own. Joining Ray on the end of the rope, they started to haul the hapless animal up the final section of the cliff.

'Think we should stand back a bit,' I said to Brian, 'or else we'll get a shower of rocks on our heads.'

Once more Brian grunted something to himself and did quite the opposite of what I'd suggested.

'I'll give it a hand up,' he said as, with only one hand holding the rope, he was on his way again, this time upwards to where the lamb had now become caught under an overhanging rock.

With a push from Brian's outstretched hand the lamb was soon on its way again. Two more shoves and a lot of pulling from the top and the lamb at last disappeared over the upper edge of the cliff and out of my view. A second later and Brian followed. Success, and what a performance, I thought to myself.

Until now I'd envisaged going back up by way of the path on which I'd come down, but it dawned on me that, having watched Brian scale the heights above one-handed, there was no way that I could take the easy route. Once more adopting the 'don't look down' approach, I reluctantly began the climb.

The next minute or two were nerve-racking in the extreme. It was a mighty relief when I finally hauled myself back over the edge and stood upright on level ground once more. Brian stood chatting to Ray, and the lamb had already run off, its night on the rocks probably forgotten already. Ray gave me a What kept you? sort of look. He didn't seem interested in a post mortem on the rescue – it had obviously been rather tame.

'When we were children, we used to collect gulls' eggs from right along this coast,' he said, pointing towards Bull Point and Morte Point, 'and we never used fancy harnesses. Come to think of it, half the time we didn't bother with ropes either.'

It wasn't a pastime I intended reviving in a hurry. As I coiled my rope I noticed that the drama had attracted a sizeable crowd of onlookers. Most were now beginning to drift away, but one lady wandered over.

'I saw the people who caused this,' she said. 'There were four of them with two dalmatians, two spaniels and a mongrel. I was down in the cove at Rockham yesterday when they chased that lamb over. They had four in the corner but it was only that one that jumped.'

It turned out she was a farmer's daughter who was on holiday in the area. She explained that she'd been so incensed she'd run up from the shore to remonstrate with the owners.

'They just shrugged and said, "It's only a sheep,"' she said, visibly angry. 'People like that shouldn't be allowed to have dogs if you ask me.'

After the days of clearing up the mess these people had made of my flock, I wouldn't have put it quite so diplomatically.

CHAPTER TWENTY

The Dogs of Windcutter Down

The sight that greeted me as the hillside market at Blackmoor Gate loomed into view was a heartening one. For the past year and more, the long lines of iron pens had been deserted, the grounds eerily devoid of any signs of life. Today the scene was transformed, with sheep filling some of the pens and a small convoy of lorries disgorging more to join them.

It had been a long hard struggle to resume Exmoor's main

livestock market. The Ministry were even now still wary of foot and mouth, and had introduced a volume of new regulations to control this aspect of the industry. A new coat of paint had been deemed necessary before the iron pens could hold sheep again, and the old, wooden ones had to be scrapped altogether. There was also a ban on livestock having contact with 'unwashable' surfaces like grass, so large areas had been concreted over. Much of the work had been undertaken by local farmers, who were determined to see their market back in full swing.

With every Ministry box ticked, Blackmoor Gate was up and running once more, but there was no way anyone could claim things were completely back to normal. As I pulled up with twenty draft ewes, it was obvious that the place was lacking the bustle of eighteen months ago. During the big autumn sales of the past, sheep had even filled row upon row of makeshift pens across the car-park, but today the throng of farmers sorting their animals almost outnumbered the sheep. Could it really be that so many animals in the locality had been wiped out?

If sheep were thin on the ground, the white-coated figures of the men from the Ministry and the Trading Standards Office certainly were not. A small army of them were patrolling officiously along the alleys, clipboards and pens poised. If form-filling were an Olympic event, then these were the men to win Britain a gold. They were all weighed down under piles of licences and other paperwork – yellow for the sellers, blue for the hauliers, white for the local authority and pink for the buyer. Not a sheep could be moved without the destruction of half a tropical rain forest, it seemed.

As if this were not enough, another team was positioned around the lorry 'wash out' station, issuing yet more forms to drivers as they hosed down their vehicles. I had to admit the sight of these particular bureaucrats still rankled with me. A few years back while at market, they'd served an enforcement notice on me. Apparently the step on to the ramp of my livestock trailer was fifteen millimetres too high.

As the morning picked up pace, a steady stream of lorries and trailers unloaded their cargoes, and slowly the market filled. By the time the auctioneer's bell rang out there were approaching a thousand animals penned for sale. A reasonable turn out, but not a patch on the ten thousand animals that the market would have coped with at its peak.

I chatted to a few seldom-seen faces before Richard caught my eye. I hadn't seen him since the skittles evening.

'Coming for a bacon roll?' he called.

'You sure that we don't need a licence for that?' I replied, hoping one of the officials might have overheard me, as we walked over towards the tin shack that served as the market café. A few minutes later we sat down with a coffee and some belated breakfast.

We spent several minutes discussing the subject that obsessed every sheep farmer, not just in the southwest but in the UK – prices. The price of finished lambs and breeding sheep depends on so many variables – supply and demand, the value of the pound and the amount of grass around, to name but a few. The truth was, no one knew whether prices would rise or fall because no one knew which way those variables were going to go. In terms of supply, the great unknown was whether those who had lost their sheep would restock, or simply turn their backs on the industry.

'Those who've farmed all their lives will go back into it,' Richard said decisively. 'It's whether the next generation start farming, that's the question.'

By now there were a dozen or more people sitting at the rickety wooden tables in the shed. I recognised many of them and knew they represented a cross section of the industry in this part of the world, from smallholders to owners of large businesses, but very few of them were under forty years of age. Richard pointed out a man with his back to me, a local livestock agent.

'John there reckons he's got over two hundred customers on Exmoor,' he explained, 'but there's less than twenty with sons wanting to carry on with livestock farming. You just think about the skills that will be lost if that happens.'

I couldn't disagree with him. I'd heard these figures too and believed, like Richard, that the lack of new farmers coming through represented the single biggest problem for the future of farming. I certainly didn't expect any of my three children to follow me into the business, despite their obvious love of the life.

'If something doesn't change soon, in twenty years' time the country won't be able to feed itself,' I said.

Sensing the conversation had taken a rather depressing turn, Richard switched tack. 'How you getting on with the displays?' he asked brightly. 'Made your fortune yet?'

'Well, I'm not sure about a fortune. We had a bit of a shaky start, but things seem to be getting better, provided that the weather plays ball, that is. We had to call one display off the other day.'

'There'll be a lot of people down for the August Bank Holiday next week, and the weather forecast is good.'

'Yes, they said that last week, and it tipped it down. I thought that now the Met Office has moved to Exeter, things would be different,' I smiled.

Richard grinned. 'Someone said to me the other week, "You'd have thought with the number of windows they've got in that building, they'd get the weather right sometimes!"'

We laughed and made our way back out to the sale, now

getting well into its stride. Things don't tend to be quite as organised at Blackmoor Gate as some other markets. I've been to sales in the north of England where by dint of masterful logistics, twenty thousand lambs can be sold through an auction ring in one day. I had to feel sorry for today's auctioneer who, microphone in hand, was taking bids for a single ewe. The sheep in question was well past her prime, but there were still two bidders interested in acquiring her. With a touch of a cap here and a twitch of an eyebrow there, they kept upstaging one another. Slowly but surely the price rose to £12.50 before the hammer finally fell. If things keep moving as slowly as this we'll be here until Christmas, I thought.

Ten minutes later the gavel was raised once more, this time on my pen of twenty ewes. This part of the sale, at least, moved along a bit more quickly. 'Going once . . . going twice . . . sold at twenty pounds,' the auctioneer was soon announcing with a tap of his gavel on his clipboard – £400 for twenty ewes: by no means a spectacular trade, but not a disaster either.

By lunchtime the cluster of buyers, sellers and market staff had worked their way through every pen and the lorries were being loaded. Like everyone else, I'd been hoping this first sale might give me a pointer as to where prices were headed. When the main autumn sale came in a few weeks' time I would have rams to buy, and a couple of hundred lambs and the remainder of the year-old ewes to sell. It was a sale that could make a big difference to the year's accounts, so any guidance I could get was valuable, but today's prices hadn't conformed to any discernible pattern. Some sheep had gone for decent prices, but others had seemed cheap on the day. As a barometer of what might lie ahead it was next to useless. Such is the unpredictable nature of farming.

Driving home in the Land Rover, however, one thing was

clear. A small part of me had headed for the market hoping that it might signal a change in fortune for the industry. A few hours later I knew that that simply wasn't going to happen overnight. Sheep farming still seemed to be in decline, perhaps a terminal one. Back down towards Mortehoe, I saw a farm had put up a new sign advertising bed and breakfast. According to the latest figures I'd read, more than 60 per cent of full-time farmers had now diversified, mostly into accommodation like this. It was just the way of things and it applied to me as much as anyone else. Branching out into new areas was the only option. If we were going to get by then I would just have to persevere with the displays, the puppies and, more's the pity, training dogs like Taffy and Trim.

I sat in the Land Rover at the top of Ivan's drive, a sudden wave of anxiety washing over me. I couldn't have been happier to be taking the troublesome Taffy home, but at the same time I knew I'd be expected to demonstrate what progress I'd made. I had, after all, been taking Ivan's money for the best part of two months.

Over the phone the previous night, I'd tried to let him down gently about Taffy's progress. 'Trim's doing pretty well,' I'd said, 'but Taffy's been a tougher nut to crack if I'm honest, Ivan. He's come on a lot, but there's still a way to go.' This was close to the truth, I had told myself ever since. Well, closeish.

I was glad to see that Ivan didn't seem too worried as he greeted me at the door. 'Come on in, the kettle's on,' he said with a cheery slap on the back. We skirted round the subject for a few minutes, exchanging words on the weather and the state of the lamb market, but it wasn't long before his curiosity got the better of him.

'I've got a few old ewes just by the house here,' he said,

draining his mug of tea. 'P'raps you could show me how Taffy's getting on.'

'Yes, of course,' I said, hoping there was at least a grain of confidence in my voice.

As we walked out to the field, I tried dampening Ivan's expectations a little more. 'He's never going to make a great dog,' I said, 'but, if nothing else, he seems to have got out of that habit of chasing off single sheep.'

There are times when you could swear that a dog knows exactly what is being said about it, even if it is out of earshot at the time. A few minutes later, with Ivan, his wife and son watching, Taffy edged towards the small flock of sheep. Even at fifty yards away, I could weigh up his mood. What with having spent an hour in the back of the Land Rover and the excitement of returning home, he was fit to burst and was obviously going to have one of his missile moments. The thought of growling at him in an attempt to stop him doing anything too explosive crossed my mind, but I reckoned that would seem unprofessional, so instead I held my breath for a second, praying that the wretched dog would remember his manners.

He didn't. When a ewe on the left-hand side of the flock drifted loose from the rest, Taffy saw his chance. With his rockets once more ignited, he was off in an instant.

'No, Taffy, no,' I shouted, but it was too late. To my horror he latched on to the wool at the back end of the sheep, then charged across the field.

'Oh,' said Ivan, looking rather surprised. 'He used to do that before you had him.' His wife shuffled about looking embarrassed, before heading off back towards the house. His son stood there speechless.

'He hasn't done that for a long time,' I said, ashamed of my excuse even as I delivered it. I tried to make amends for Taffy's bad behaviour by putting him through some of his better moves. He performed adequately enough, but the damage was done.

As I made my way through the labyrinth of Devon byways half an hour later, I felt little better than a conman. When Ivan had written out a final cheque for Taffy's training I'd known it represented no more than a kennel would have charged to feed and walk the dog for a month, but I still felt a failure. To me I'd not fulfilled the contract. I still had Trim to work with, but I'd made a decision I knew I wouldn't go back on.

'Never again,' I said to myself.

'Dad, guess what,' Clare called as I climbed out of the Land Rover in the yard. 'Jake saved a sheep's life this afternoon.'

Jake was clearly oblivious to the rather grand claims being made on his behalf and was ambling on ahead of her, with a large clump of grass in his mouth.

'Really? What happened?'

'All your sheep were in Minty's paddock, so I told Jake to chase them out for me,' Clare began.

'I didn't know that he would work for you,' I interrupted, a little surprised.

'Well, he doesn't really work for me, it's just that he usually rounds up the sheep by himself, whenever we walk through them.'

I'd been trying to encourage Clare to take an interest in working with the sheepdogs for a while now, so was pleased by this.

'Does he come back to you when you call him?'

'He's fine, Dad,' she said, waving away my overly detailed enquiries about these impromptu gatherings. 'Anyway, earlier on, when he went behind that old brick tank in Minty's paddock, he didn't come out for a long time. When I called him he came, but he kept looking back towards it, so I went to see what he was looking at.'

'And . . .' I said, now thoroughly intrigued by her story.

'And there was a lamb with its head stuck in the fence,' she said triumphantly. 'I'd never have seen it without Jake. He wouldn't leave it behind.'

I was impressed – by both of them.

'Well done, that was very clever of both of you. Didn't it feel good to be working like that with your own sheepdog?' I asked after a pause, hoping that the incident may have lit a little spark of interest.

'Not really,' Clare said dismissively, her mind already wandering off elsewhere.

The incident only confirmed my opinion of Jake and underlined why I was feeling so delighted with his progress. He had developed what seemed to be two distinct modes – work and play. While with Clare or on any other 'social' outing he would bound along, with any convenient object fixed in his mouth, flicking molehills into the air with his nose. In short, he behaved like a proper fool, which it had to be said rather endeared him to his human owners. His laidback approach to life seemed to endear him to his kennel mates too. Greg, in particular, might have had a problem in accepting a young pretender into his kennel, but seemed to have taken to Jake without feeling that he was a threat to his 'top dog' status.

However, it was his work mode that was the most encouraging for me. For all his odd behaviour, Jake was showing fantastic progress as a working sheepdog. Something in his head seemed to switch on the moment he got a chance to interact with sheep, and his whole demeanour was different from other young dogs when this happened. He seemed to weigh up every situation, then respond to it sensibly. In addition, he was always listening to my voice, so much so that a misplaced cough would make him stop and look around, as if he thought that he'd missed an instruction. Jake had a level of intelligence that I hadn't seen in one of my dogs since the early days of Greg. Secretly, I was beginning to get quite excited about the prospect of taking him to sheepdog trials in years to come.

'You'll be able to claim that you did some of his training, when he starts winning trials,' I told Clare.

Debbie had just arrived from the house, carrying a boxful of fliers towards the car. 'You can't take him to a sheepdog trial,' she called, catching the end of the conversation. 'They'll all laugh at you, trying to compete with a collie crossed with a dalmatian.'

'Block your ears, Jake!' I called back as the great lolloping hound bounced over to me. 'She doesn't know what she's saying.' I turned to Debbie, 'Of course he'll be good enough for trialling.'

Debbie took another look at the white and black spotted dog, who now stood staring at a manhole cover in the middle of the yard. He was poised to pounce on it, a habit he'd developed over the last few weeks, none of us knew why.

'Don't say I didn't warn you,' she said as Jake leapt into the air, bounced on the steel cover, then ran full speed in a great circle around the yard.

'As my Aunt Jill used to say, "Every wedding needs an eccentric,"' I said defiantly.

Late August can throw up an unwelcome variety of weather in our little corner of North Devon, everything from thunderstorms and driving rain to long, dank days of blinding fog, but today, fortunately, the sky was set fair with not a hint of change to come. The cawing of a pair of ravens echoed up

Borough Valley and a fresh breeze rustled the gold-tinged leaves in the uppermost reaches of the beech trees, offering the first inkling that autumn wasn't too far away now.

'Well, at least we're not going to get rained off for the last display of the year,' Debbie said as I loaded the usual collection of display boards, direction arrows and bales of straw into the trailer.

'There seem to be plenty of people about,' I said, in a more positive frame of mind than her for once. 'Perhaps you'd better get up to the field in plenty of time this afternoon, just in case there's a good turn out.'

Wednesday afternoons had taken on a routine of their own. Up on Windcutter Down, a dozen compliant ewes had to be put in position, and the various pens and gates I'd strategically placed around the field had to be stood up after being flattened by the sheep that had spent the week grazing there. The most time-consuming job was usually placing a flock of suitable sheep in the farthest field, half a mile away from the audience.

For me, the gathering of the flock from such a long distance was the ultimate demonstration of the real capabilities of the dogs. I'd discovered that this was the best part of the evening for the audience as well, which pleased me. The only headache was that I needed at least a couple of hundred ewes for this climax of the display.

It was important to make sure the sheep were looking their best. The sight of a substandard animal hanging to the back of the flock would look poor in front of the visitors, so a couple of hours beforehand, I ran the selected sheep through the handling pens back at the farm. Now, as I passed my eye over them, I thought they looked in pretty good order. Indeed the whole flock had thrived on the dry conditions of late July and August. It was good to feel that the endless days of sorting, foot-trimming, foot bathing and docking had been to some avail.

If only every other aspect of the farming business had been in such near-perfect shape, I thought as I admired the sheep for a moment. Sales had been disappointing this summer. The new lamb crop had continued to grow in size, but the lambs had failed to put on the necessary flesh in order to sell as 'finished' stock. I'd have to sell some as 'store' lambs for further fattening once the autumn sales started in earnest, but for the time being the lack of income had left us starved of funds again. It was yet another reminder of the knife-edge that our farming business had been poised on for what seemed an eternity.

However, unlike a year ago, at least I knew the lambs would sell soon enough, and a good turn out for the last display of the year would set us on our way towards the winter in a more positive frame of mind than we'd been for a long time.

There was still warmth in the late afternoon sun when Debbie arrived to take up her position on the entry gate where she was joined by Clare, wearing a newly sewn money belt. Nick and Laura arrived on bikes and cycled around without a care in the world. All three children were mercifully oblivious to the reason for these Wednesday evenings. Or at least, I hoped they were. Deep down they probably felt the whole enterprise was born of some need for their father to show off his sheepdogs. It hadn't occurred to any of them that sheep farming could no longer pay the bills.

It was nearly an hour before the display, but there were already two cars parked in the field. It seemed that most vital of advertising elements – word of mouth – had begun to work in our favour. 'The people in the next door caravan came last week, and said how good it was, so we thought we'd come and have a look,' one car's occupants told me. Whether they were just trying to be nice or not, I didn't really care. It was just what I needed to hear.

By twenty to six, Debbie and Clare were being pressed into serious action. It began with the odd car, but within a few minutes, the trickle became a constant stream. For the first time a queue formed on the drive, forcing Debbie and Clare to split up and work their way along the line of cars and the outstretched hands with money that appeared through the windows.

My preparations were complete, so all I had to do for the next few minutes was watch the audience arrive. The old worry of no one turning up was slowly being replaced by a different concern. How was I going to cope with performing in front of so many people?

At least I could console myself with the thought they'd have somewhere to sit. The row of straw bales I'd been setting around the working arena until now provided seating for fifty or more people, but over the last week I'd decided the time had come to put in a second, higher 'back row'. I'd achieved this by upturning the feed troughs I used in the sheep shed during the winter. Their wooden undersides made a perfect 'perching' seat – once the worst of the winter's dung had been brushed off, that was.

The back row was soon full to capacity, and it was standing room only for the groups that now picked their way through the rows of parked cars. As six o'clock arrived, vehicles were still flowing into the field.

'I don't know what you've done with the advertising this week,' I called to Debbie, 'but it's certainly worked.'

She looked at me anxiously. 'I'm not sure if we can cope with too many more people; they won't all be able to see,' she said.

I looked up the drive. Although the stream had slowed

down, there were still four or five cars waiting in Clare's queue.

'I'd better make a start,' I told her.

Slowly I'd learned what interested the audience, and occasionally what made them laugh, but as I launched into my by now well-rehearsed spiel, I realised I was looking out at what appeared to be more than two hundred faces. I swallowed hard.

If the nerves were still obvious in my voice, there was no sign of stage fright in the dogs. Greg, Swift, Fern, Gail and Ernie each performed their part of the show without a single hitch. In fact, the early part of the evening passed off so well that half an hour into the display I decided to take a chance.

'This is Jake, the latest recruit to the kennel,' I said, simultaneously attaching a lead to his collar as he leapt from the Land Rover.

My confidence was temporarily dented by the sound of someone exclaiming in a loud voice, 'That's not a sheepdog,' but by now I was used to such comments and took them as a challenge. I proceeded to give Jake a great build-up, telling everyone that he was the cleverest dog that I had had since Greg was a puppy.

'Jake is only eleven months old, but he's already capable of collecting those sheep from the other side of the field,' I said, gesturing at the packet of ewes I'd just used with the other dogs.

The other side of this field was only a matter of a hundred and fifty yards, a distance that Jake could easily cover, but, in the brief few moments it had taken me to get him in position, the small flock had starting moving away, and by now they were slowly gaining pace and heading towards the gap in the wall between two fields.

All of a sudden, I could see what was going to happen. If I

didn't stop talking and get on with things, the sheep were going to disappear from view. With a quick 'Come bye, Jake' I set him free from his lead.

By the time he was on his way, the sheep had reached the gap in the wall and were gone. Jake, however, was undeterred. As the surprised audience glanced first towards the dog, then towards me, presumably looking for signs of panic, Jake rushed onwards, through the same gap. Then he too was lost from view.

I didn't have to look around to know there was a lot of murmuring and pointing going on behind me, but, somehow, even though I had a pup in an awkward situation, the sense of panic I often got was missing. As the seconds slipped by, the murmuring became a hubbub of chatter, but just as I was beginning to wonder if my faith had been misplaced, seven ewes trotted back into view. Behind them, his distinctive white and black features unmistakable in the evening sun, was Jake, looking fully in control.

A burst of applause erupted from behind me. Jake seemed to play to the crowd, bringing the sheep closer, and turning on the style. 'The dog's a natural,' I said to myself.

When the display came to an end twenty minutes later, a crowd of admirers thronged around the dogs, Swift as usual accepting the adoration of at least a dozen children, a role that she seemed to love.

A man about my own age pulled me to one side. 'I really enjoyed your show,' he said introducing himself as Jonathon. 'I'm a professional falconer, perhaps we could do some shows together next year?'

We talked about the idea for a few minutes. It seemed to make a lot of sense. 'Give me your number and we'll get something organised,' I told him.

The last car left the field just as the light began to fade from the evening sky. Far down in the valley, where the shadows were already lengthening into dusk, a thousand rooks gathered in a noisy mass, swirling towards the tree tops. Debbie wore a satisfied grin as she produced a flask of coffee. She hadn't counted the proceeds of the evening's work, but it had obviously been well worthwhile. 'That's the best crowd we've had in. Well over two hundred and fifty people, I'd say. We've probably made twice as much from the displays this summer as you made from shearing last year. It's got to be the way ahead.'

'Especially if we add to it with a bit of falconry,' I said, taking the opportunity to fill her in on the offer I'd had from Jonathon.

As Debbie poured us a couple of mugs, Clare, Laura and Nick busied themselves by dragging the straw bales back towards the trailer, staggering as they did so.

Greg pricked his ears and ambled over wagging his tail, exuding happiness. 'You're a good lad, Greg,' I said, rubbing his chest.

He looked at Debbie and back towards me, then, not knowing quite what was expected of him, lifted his paw and placed it on my knee.

His moment with the boss didn't last long. Soon, one by one, the rest of the dogs were wandering over to join us. 'And I bet you don't know how important to us you lot have just become,' I said, as I gave Swift, Gail, Fern, Ernie and Jake a ruffle as well.

List of Illustrations